All Solid State Thin-Film Lithium-Ion Batteries
Materials, Technology, and Diagnostics

T0225559

Tatiana Kulova
Frumkin Institute of Physical Chemistry and
Electrochemistry of the Russian Academy of Science
Moscow, Russian Federation

Alexander Mironenko
Institute of Physics and Technology of Russian
Academy of Science, Yaroslavl Branch
Yaroslavl, Russian Federation

Alexander Rudy
Institute of Physics and Technology of Russian
Academy of Science, Yaroslavl Branch
Yaroslavl, Russian Federation

Alexander Skundin
Frumkin Institute of Physical Chemistry and
Electrochemistry of the Russian Academy of Science
Moscow, Russian Federation

CRC Press
Taylor & Francis Group
Boca Raton   London   New York

CRC Press is an imprint of the
Taylor & Francis Group, an **informa** business
A SCIENCE PUBLISHERS BOOK

Cover credit: The image used on the cover has been made by A. Rudy.

First edition published 2021
by CRC Press
6000 Broken Sound Parkway NW, Suite 300, Boca Raton, FL 33487-2742

and by CRC Press
2 Park Square, Milton Park, Abingdon, Oxon, OX14 4RN

Library of Congress Cataloging-in-Publication Data

Names: Kulova, Tatiana, 1962- author.
Title: All solid state thin-film lithium-ion batteries : materials,
    technology, and diagnostics / Tatiana Kulova, Frumkin Institute of
    Physical Chemistry and Electrochemistry of the Russian Academy of
    Science, Moscow, Russian Federation, Alexander Mironenko, Institute of
    Physics and Technology of Russian Academy of Science, Yaroslavl Branch,
    Yaroslavl, Russian Federation, Alexander Rudy, Institute of Physics and
    Technology of Russian Academy of Science, Yaroslavl Branch, Yaroslavl,
    Russian Federation, Alexander Skundin, Frumkin Institute of Physical
    Chemistry and Electrochemistry of the Russian Academy of Science,
    Moscow, Russian Federation.
Description: First edition. | Boca Raton : CRC Press, 2021. | Includes
    bibliographical references and index.
Identifiers: LCCN 2021000986 | ISBN 9780367086824 (hardcover)
Subjects: LCSH: Lithium ion batteries. | Solid state batteries.
Classification: LCC TK2945.L58 K85 2021 | DDC 621.31/242--dc23
LC record available at https://lccn.loc.gov/2021000986

ISBN: 978-0-367-08682-4 (hbk)
ISBN: 978-0-367-76343-5 (pbk)
ISBN: 978-0-429-02373-6 (ebk)

Typeset in Palatino Roman
by Innovative Processors

Acknowledgments

This work was financially supported by the Ministry of Science and Higher Education of the Russian Federation within the framework of the state task of Frumkin Institute of Physical Chemistry and Electrochemistry of the RAS (Chapters 1 and 2), and within the framework of the state task of P.G. Demidov Yaroslavl State University No. 0856-2020-0006 (Chapters 3 and 4).

Preface

In 2019 the Nobel Prize in Chemistry was awarded to John B. Goodenough, M. Stanley Whittingham and Akira Yoshino for creating lithium-ion batteries which was a brilliant demonstration of the huge importance of such batteries for modern civilization. Indeed the announcement on launching the production of lithium-ion batteries in 1990 marked a revolution in the field of rechargeable batteries. Lithium-ion batteries became a base for modern portable electronics, including cellular phones, notebooks, digital cameras, wireless power tools etc. The annual world output of lithium-ion batteries amounts to several billion products. Lithium-ion batteries are featured with high energy density, high operating voltages, low self-discharge, low maintenance requirements and wide range of operational temperatures.

In the more recent years, great efforts have been directed towards the development of scaled-up lithium-ion battery systems, in particular for electric vehicles (or at least hybrid electric vehicles and plug-in electric vehicles), renewable energy systems and smart grids.

All-solid-state thin-film lithium-ion battery presents a special and very important version of such devices. Indeed all-solid-state batteries possess certain advantages compared with common lithium-ion batteries based on a liquid electrolyte. First, the absence of organic solvents increases intrinsic safety of the battery due to eliminating the risk of possible fluid and vapor leakage and therefore diminishing the risk of fire and explosions. Second, the solvents of the liquid electrolytes often participate in the degradation processes of common lithium-ion batteries, which is why the cycle-life performance of solid-state batteries is supposed to be much better. Third, the use of liquid electrolyte results in several restrictions for the battery's design and size. (Typical thickness of common separators in lithium-ion batteries is about 20 μm, whereas the thickness of solid electrolytes amounts to 1 μm.) Therefore, all-solid-state concept opens the way to creation of microbatteries.

The need for all-solid-state thin-film lithium-ion batteries appears from rapidly developing microelectronics, especially battery-powered integrated circuit cards (smart-cards), radio-frequency identifier (RFID) tags, smart watches, implantable medical devices, remote microsensors and transmitters, Internet of Things (IoT) systems and various other wireless devices including smart building control and so on. Often these batteries must be placed on the same crystal (chip) as the microelectronics device itself, as a so-called embedded system. The all-solid-state thin-film lithium-ion batteries manufacturing technology should be compatible with the manufacturing technology of an integrated microcircuit, an MEMS device, a semiconductor sensor, etc., i.e., in general, it must be VLSI-compatible. Rather important kinds of thin-film batteries are flexible and transparent ones. It is worth noting that recently, a significant progress for all-solid-state lithium-ion batteries has been made by experimentally developing and optimizing solid electrolytes and the electrode functional materials.

The present monograph provides an overview of developments in the field of materials, technologies and diagnostics of integrated lithium-ion batteries. The monograph sets out the basic principles of the operation of thin-film all-solid-state lithium-ion batteries and their electrochemical systems, the characteristics of electrode materials and solid electrolytes are given. The methods of applying (for deposition) of the functional layers of thin-film lithium-ion batteries, the methods of manufacturing test structures and the testing methods of lithium-ion batteries and their individual elements are considered. A description is made of the methods for diagnosing (diagnostics) thin-film lithium-ion batteries materials and the features of their investigation by Scanning Electron Microscopy, Characteristic X-ray Spectroscopy and Mapping, X-ray Crystallography and Phase Analysis, Electrochemical Impedance Spectroscopy.

The monograph is intended for teachers, researchers working in the fields of: materials science, electrochemistry, thin-film physics and technology, methods of vacuum deposition and diagnostics of materials for lithium-ion batteries, for graduate students and university students, as well as for developers and manufacturers of thin-film lithium-ion batteries. One of the goals of the monograph is to assist in the training of specialists in the field of electrochemistry, semiconductor physics, electronics and circuitry, mathematical modeling, thin-film and MEMS technologies, and serve as the basis for communication among members of interdisciplinary scientific groups. To what extent the authors have managed to achieve this goal, could be judged by the readers.

The authors of the monograph are experts in fundamental electrochemistry, theory of chemical power sources (especially in lithium-ion batteries) on the one hand, and in the theoretical physics, microelectronics, semiconductor technologies on the other.

The monograph consists of four principal chapters. Chapter 1 presents fundamentals of rechargeable batteries, as well as features of lithium-ion batteries and especially of thin-film batteries. Chapter 2 is devoted to materials for electrodes and electrolyte of all-solid-state lithium-ion batteries. Chapter 3 describes technological aspects of applying (deposition) functional layers of all-solid-state lithium-ion battery. Chapter 4 is concerned with the problems of diagnostics of functional layers of all-solid-state lithium-ion battery. Chapters 1 and 2 have been written by A.M. Skundin and T.L. Kulova, Chapters 3 and 4 have been written by A.S. Rudy and A.A. Mironenko.

The authors will be very grateful for criticism, notes and suggestions.

Tatiana Kulova
Alexander Mironenko
Alexander Rudy
Alexander Skundin

Abbreviations and Acronyms

AC	Alternating Current
ADPESS	A-Site-Deficient Perovskite Solid Solutions
AM	Additive Manufacturing
BSE	BackScattered Electrons
CAD	Computer Aided Design
CEC	Chemical-Electrochemical-Chemical
CPE	Constant Phase Element
CVD	Chemical Vapor Deposition
CXA	Computer X-ray Analyzer
CXR	Characteristic X-ray Radiation
DC	Direct Current
DIW	Direct Ink Writing
DoD	Depth of Discharge
DPG	Diphenyl Germanium
EDA	Energy Dispersive Analysis
EDS	Energy Dispersive Spectrometer
FDM	Fused Deposition Modeling
FEG	Field Emission-Guns
FWHM	Full Width at Half Maximum
IC	Integrated Circuit
ICDD	The International Center for Diffraction Data
IoT	Internet of Things
IUPAC	International Union of Pure and Applied Chemistry
LAGP	Lithium-Aluminum-Germanium-Phosphate
LATP	Lithium-Aluminum-Titanium-Phosphate
LCD	Liquid Crystal Display
LED	Light-Emitting Diode
LiPON	Lithium Phosphorus Oxynitride
LiPOS	Lithium-Phosphorus-Oxygen-Sulfur
LISICON	Li-Super Ionic CONductor
LiSON	Lithium-Sulfur-Oxygen-Nitrogen
LLT	Lithium-Lanthanum-Titanium
LLTO	Lithium-Lanthanum-Titanium-Oxide
LTO	Lithium-Titanium Oxide
MEMS	Microelectromechanical System

NASICON	Na-Super Ionic CONductor
NCA	Nickel-Cobalt-Aluminum
NIST	National Institute of Standards and Technology
NLO	Nickel-rich Layered Oxides
NMC	Nickel-Manganese-Cobalt
OCV	Open Circuit Voltage
ORNL	Oak Ridge National Laboratory
PDF	The Powder Diffraction File
PECVD	Plasma Enhanced Chemical Vapor Deposition
PVD	Physical Vacuum Deposition
RF	Radio Frequency
RFID	Radio-Frequency Identifier
SCA	Single-Channel Analyzer
SCR	Space Charge Region
SDD	Silicon Drift Detector
SE	Secondary Electrons
SEI	Solid Electrolyte Interphase
SEM	Scanning Electron Microscopy
SLA	Start, Light, Ignition
SLI	Solution Liquid Solid
SLS	Stereolithography
SRAM	Static Random-Access Memory
SRM	Standard Reference Material
SSD-BSE	Solid-State Detector of BackScattered Electrons
SSLIB	Solid-State Lithium-Ion Batteries
UV	Ultra-Violet
VLS	Vapor–Liquid–Solid
VLSI	Very Large-Scale Integration
WDA	Wave Dispersive Analysis
WDS	Wave Dispersive Spectrometer
XPA	X-ray Phase Analysis
XSA	X-ray Structure Analysis

Contents

Introduction

Currently, a lithium-ion battery is the most common type of battery, which is widely used as a power source in consumer electronics, electric vehicles and energy storage devices in energy systems. Thanks to lithium-ion batteries and their high capacity, it became possible to create devices such as cell phones, laptops, digital cameras, camcorders and even electric cars. The high specific capacity of the battery allows in developing relatively small power sources and chargers (power banks) that ensure the operation of mobile and portable electronics for a fairly long time.

However, over the past two decades, many new microelectronic devices and microelectromechanical system (MEMS) have appeared, for which conventional lithium-ion batteries have proved to be unsuitable due to their size. These are smart cards, RFID tags, MEMS devices, flexible electronic devices, implants and stimulators, microrobots, etc. These devices require batteries in microminiature, hence in integral implementation. Conventional lithium-ion batteries, consisting of active layers two hundred micrometers thick, separated by an electrolyte-impregnated separator, are not applicable here. It is not only due to dimensions, but also the technology of their manufacture, based on the electrodes coating with paste-like material with subsequent baking and pressing, cutting, stacking, filling with liquid electrolyte, etc.

The fact is that, under operating conditions, SSLIBs must withstand high gravitational overloads and operate in a wide pressure range from tens of atmospheres to high vacuum. Often, these batteries could be formed on the same chip as the integral circuit itself or MEMS. In this case, the SSLIB manufacturing technology should be compatible with the manufacturing technology of the microcircuits, MEMS devices, semiconductor sensors, etc. It follows from the above that the SSLIB manufacturing technology in the general case should be integral, and in certain cases VLSI-compatible.

Therefore SSLIBs are fabricated using PVD-deposition methods, making the basis of microelectronics technology. In the manufacture of 3D SSLIBs, plasmochemical etching methods are additionally used to form the relief. In terms of technology, SSLIB is a common microelectronic device, the same as an integrated circuit. Its only difference from the microcircuit is the lack of topology and a large surface area. The technologies and materials used for its manufacture are also standard technologies and materials (except solid electrolyte) for microelectronics. And the last production operations, making the SSLIB's manufacturing route, consist of sequential PVD-deposition of functional layers of the battery and differ from the production of integrated circuits only in significantly fewer operations.

At the stage of SSLIBs technology development, often especially in the characterization of the structure formed by functional layers, and their elemental and phase composition analysis are required, as well as mass and charge transfer processes in functional layers and across interlayer boundaries. To diagnose both SSLIBs materials and the batteries themselves, basically the same methods are used as for microelectronic devices. First, this is scanning electron microscopy, X-ray microanalysis, X-ray structural and X-ray phase analysis and impedance spectroscopy. The particular kind of analysis here is only impedance spectroscopy, or rather its frequency range, not typical for integrated circuits. For SSLIB materials and devices based on them, a typical frequency range of measurements makes $10^{-3} - 10^6$ Hz.

The SSLIBs' features listed above determine the structure of the monograph and the content of its sections. In the first chapter, the fundamentals of rechargeable batteries, including general principles of their functioning, common problems of design and performances are considered along with various conventional kinds of rechargeable batteries. The main part of the first chapter is devoted to lithium-ion batteries. The chapter also describes features of thin-film batteries. In the second chapter, materials responsible for solid-state thin-film lithium-ion batteries, the most promising functional materials of positive and negative electrodes, as well as materials of solid electrolytes, are examined in detail. Both the basic properties of materials and the methods of their synthesis, and briefly the methods of their use in solid-state batteries are considered. The third chapter describes the PVD methods used in SSLIBs manufacturing. The fourth chapter is devoted to the description of the basic methods for diagnostics of SSLIB structure, elemental and phase composition of its functional layers and lithium transfer mechanisms.

The monograph is written within an interdisciplinary approach and is intended primarily for teachers and students of chemical and physical faculties, specializing in the field of lithium-ion batteries in general and SSLIB in particular. For this purpose, a number of topics are presented at the most accessible level and provided with relevant explanations and examples. The book will be most useful for graduate students, studying research in related fields of science and technology at the intersection of electrochemistry and microelectronics. The book may also be helpful to the general reader who is interested in the problems of modern energetics and wants to get an idea of the direction and level of developments in the field of SSLIB.

1

Modern Lithium and Lithium-Ion Rechargeable Batteries

1.1 Fundamentals of rechargeable batteries.

1.1.1 General principles

Definitions and working principles. Rechargeable batteries concern a wide class of devices with the common name 'chemical power sources' [1]. Any chemical power source represents a device for direct conversion of chemical energy into an electrical one. More strictly, 'chemical energy' is energy of a chemical reaction occurring in a power source. This reaction is called a 'current-producing reaction'. For instance, interaction of lithium metal with manganese dioxide is a current-producing reaction in popular button cells responsible for watches, remote controllers, calculators and so on. This reaction can be expressed by the equation

$$Li + MnO_2 \rightarrow LiMnO_2 \tag{1.1}$$

Every current-producing reaction is a redox-process, i.e., the interaction between the reducer and oxidizer. On oxidizing a reducer releases electrons whereas an oxidizer gains electrons at reduction. For the process (1.1) MnO_2 is the oxidizer, and manganese valency changes in this case from +4 to +3. Lithium metal is a reducer, and its valency changes from 0 to +1.

Chemical power source consists of one or several galvanic cells. It is a galvanic cell in which the conversion of chemical energy in to an electrical one occurs.

The salient feature of a current-producing reaction is that it consists of spatially divided processes of reduction and oxidation. Each of these processes take place at the interface between an electrode and an electrolyte. In any case, the electrodes are made from materials with electronic conductivity, so-called first-class conductors (metals, carbonaceous materials, for instance). The electrolyte is the material with ionic conductivity, the so-called second-class conductors (solutions, melts and so on). In the above example the reduction process is expressed by the equation

$$MnO_2 + Li^+ + e^- \rightarrow LiMnO_2 \tag{1.2}$$

The oxidation process is expressed by

$$Li \rightarrow Li^+ + e^- \tag{1.3}$$

It is easy to make sure that summation of Equations (1.2) and (1.3) gives the equation (1.1).

Therefore, every galvanic cell (or simply, cell) consists of two electrodes divided by an electrolyte. When functioning (discharge) of the above-mentioned cell the lithium ions are extracted from lithium electrode into the electrolyte, then transferred by the electrolyte to another electrode, where they interact with manganese oxide. The electrons at that time are transferred from a lithium electrode to manganese-dioxide-electrode by an external electrical circuit, producing useful work. That is why, partial reactions in chemical power source are always coupled ones. In the other words, the rate of electrons' release at one electrode is strictly equal to that of electrons' gain at another electrode.

The schematic of galvanic cell and the main processes in it are depicted in Fig. 1.1.

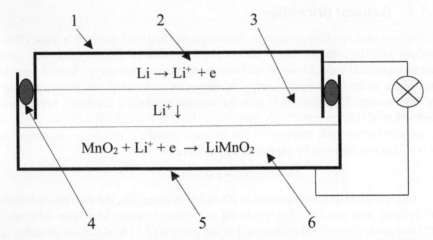

Figure 1.1. Schematic of typical galvanic cell: (1) cover, negative terminal, (2) negative electrode, (3) separator impregnated with an electrolyte, (4) insulator, (5) case, positive terminal, (6) positive electrode

It is worth noting that spatial division of partial processes of reduction and oxidation is an indispensable condition for producing electrical energy. If the reaction (1.1) is carried out in a flask by providing intimate contact between manganese dioxide and lithium, no electrical energy will be produced even though the electrons will transfer from lithium to manganese dioxide. In such situation the electron transfers will be spatially random and the total energy of a chemical reaction will be released in the form of heat.

The electrode at which oxidation occurs is referred to as 'anode'. Its counterpart at which reduction occurs is called 'cathode'. In other words, the anode is an electrode through which the electric current flows from the external circuit into the electrolyte, whereas the cathode is an electrode with an opposite current direction. If two different electrodes are divided by an electrolyte, a potential difference develops between them. If no current flows in the cell, the potential of the electrode containing a reducer is more negative because it has a higher tendency to release electrons. In the above-mentioned cell Li is a negative electrode, and manganese dioxide is a positive one. In the course of discharge a negative electrode is the anode, and the positive electrode is the cathode.

The combination of reactants of the current-producing reaction, namely the reducer and oxidizer, as well the electrolyte forms the electrochemical system of a given chemical power source. Conventionally, an electrochemical system is written as

$$(-) \text{ reducer } | \text{ electrolyte } | \text{ oxidizer } (+)$$

(Short vertical bars denote electrode-electrolyte interfaces, in which the electrochemical reactions like (1.2) and (1.3) occur). The electrochemical system of the above-mentioned cell can be expressed as

$$(-)Li|LiClO_4|MnO_2(+)$$

The electrolyte in this case is the solution of $LiClO_4$ in propylene carbonate. (The solvent is usually omitted in the notation of an electrochemical system).

According to Faraday's laws, the total charge passing through an electrochemical cell is unambiguously related to the quantity of the reactants consumed in the electrochemical reaction, and to the quantity of the products of the reaction. In the other words, the Faraday's laws can be expressed by the equation

$$Q = nFm/A \qquad (1.4)$$

Here Q is a charge passing through an electrochemical cell, m is weight (mass) of reactant, A is its molecular (atomic) weight (mass), n is the number of electrons participating in the electrochemical process and F is a fundamental constant, so-called 'Faraday number' (or simply 'Faraday') equal to 96.487C/mole or 26.8 Ah/mole. The charge nF, therefore, corresponds to conversion of 1 mole of reactant. In reaction (1.1), the oxidation of 1 mole of lithium (7 g), the reduction of 1 mole MnO_2 (87 g), and the formation of 1 mole of $LiMnO_2$ (94 g) are accompanied by the transference of 26.8 Ah. The Faraday's laws are, in essence, the conservation law in electrochemistry.

The important consequence from Faraday's laws is the fact that current I, passing for the time t through a galvanic cell, is directly proportional to the rate of electrode reactions, v, as well as to the rate of overall current-producing reaction. In other words, the cell current is a measure of reaction rates in electrical units:

$$I = dQ/dt = nFv = nF(dm/dt)/A \qquad (1.5)$$

Faraday's laws give a possibility to calculate the amount of the reactant, necessary for producing a unit charge. This quantity is called 'theoretical specific consumption of a reactant' g^T. For instance, the specific consumption of Li is equal to 7/26.8 = 0.261 g/Ah. The reversal value, i.e., the charge which can be produced by the unit weight of a reactant is called 'specific capacity' q. For lithium this quantity is equal to 26.8/7 = 3.83 Ah/g. In real galvanic cells, specific consumption g is always more than the theoretical value. There are different reasons for such a situation. The reactant cannot be fully consumed to the end of a cell discharge or it can be wasted in any side reactions and so on. The ratio $\lambda = g^T/g$ is called the utilization coefficient of a given reactant. This coefficient depends on the nature of the reactant, cell design, operation mode, and it may vary in a rather wide range from 0.3 to 0.99.

The current-producing reactions have place in the boundary (interface) between an electrode and electrolyte. Obviously, the more the area of this interface, the more will be the current I. That is why, a quotient of current to surface area of electrode (S), named 'current density' i is used as a measure of the rate of the electrode process

$$i \equiv I/S \tag{1.6}$$

It is convenient and natural to express a voltage of a cell U as the difference of potentials of positive and negative electrodes (E^+ and E^-).

$$U = E^+ - E^- \tag{1.7}$$

Obviously, a cell's voltage is always a positive value. In the opposite case the chemical power sources cannot produce the electric energy. The origin and physical meaning of the potential are rather difficult in understanding and will be commented on later. In the situation when no current flows in the external circuit (and therefore, in the cell per se) the voltage has the name 'Open-Circuit Voltage' (OCV). The OCV (U_{oc}) as well as the potentials of individual electrodes in open-circuit conditions depend exclusively on the chemical nature of the electrodes and electrolyte, and to a lesser extent on temperature, but not on the sizes of electrodes and a cell as whole.

Classification: The principles of functioning chemical power sources are classified as follows.

(1) Primary (single-discharge cells). Usually, these very devices are called 'galvanic cells', or simply 'cells'. A primary cell contains a certain limited amount of reactants participating in the current-producing reaction. Primary cells are not intended to be recharged, so once this amount of reactants is consumed, such cells cannot be used again.

(2) Secondary (storage or rechargeable) batteries. After complete or partial discharge, such a battery can be recharged by forcing an electric current through the battery in the reverse direction. Such a procedure provides regeneration of the initial reactants from the products of current-producing reaction. In the secondary battery the electrical energy from external power sources, for instance from an electric grid, is converted into chemical energy. Therefore, a secondary battery functions in some cyclic mode: (a) discharge phase, when the current-producing reaction occurs, and electric energy is delivered to a consumer, and (b) charge phase (electrolysis), when the overall current-producing reaction occurs in the opposite direction. These reactions in the secondary battery inevitably must be chemically reversible. Modern secondary batteries are able to sustain a lot of charge-discharge cycles (hundreds and even thousands).

(3) Fuel cells. The fuel cell is a primary cell, where reactants of current-producing reaction are not in the cell body but are stored elsewhere and continuously fed into the cell, while the reaction products are continuously removed. That is why, a fuel cell can continuously deliver electric energy for a considerably long time.

Such a classification is rather simplified. Indeed, in certain conditions some primary cells could be recharged, although the number of such charge-discharge cycles is limited. Secondary batteries in some cases are used as primary ones for a single discharge.

Some special kinds of primary cells are presented the by so-called reserve cells. Reserve cells are based on very active reactants, which cannot be kept in direct contact with electrolytes over a sufficiently long time. The reserve cells are stored in an inactive state and must be activated immediately before the discharge. There are two classes of reserve cells. In one, the cell does not contain any electrolyte, and a liquid electrolyte is injected into the cell just before the discharge. In another category, the cell contains an electrolyte in a solid (therefore nonconductive) state. Just before discharge, the electrolyte is rapidly heated to melting point and becomes a good ionic conductor.

Supercapacitors are very close to usual rechargeable batteries in which energy storage consists in charging of electrical double layers. Of late, some combinations of double layer capacitors and secondary batteries, so-called hybrid (asymmetrical) supercapacitors have become very popular.

In addition to classification by the principle of functioning, other methods of classification are also used, for example, according to shape (cylindrical, prismatic, disk (coin or button) cells, as well as thin film and flexible cells), electrolyte type (aqueous, including alkaline, acidic, or neutral, nonaqueous, molten, solid, polymer electrolyte), working temperature (ambient-temperature, intermediate-temperature, high temperature), application (stationary or portable cells), size (miniature, small-, medium- or large-sized) and the way of electric power generation (low-power or high-power cells).

Performances. As mentioned above, the OCV of power source first of all depends on the chemical nature of the electrodes and electrolyte, i.e., on the electrochemical system. Besides, it is somewhat affected by different factors, including the electrolyte concentration, gas pressure (for the cells with a gaseous reactant), degree of discharge, temperature and so on. In a few cases, the OCV is a fairly reproducible quantity, provided the above parameters are fixed.

When a current flow through a galvanic cell, the discharge voltage (U_d) is reduced, and the higher the current the lower the discharge voltage. At the charge of secondary battery, the charge voltage (U_c) grows with increase of the charge current. In a simplified way, two factors determine change in voltage in on-load conditions, namely so-called ohmic drop ΔE_{ohm} connected with current flow through inner resistance R

$$\Delta E_{ohm} = IR_{in} \tag{1.8}$$

and electrode polarization (overpotential) η, i.e., shift of electrode potential on going from open circuit conditions to conditions at current I

$$\eta = |E^{(I)} - E^{(0)}| \tag{1.9}$$

The causes of polarization will be described later.

Therefore at discharge of power sources

$$U_d = U_{oc} - \eta_+ - \eta_- - \Delta E_{ohm} \qquad (1.10a)$$

and at charge of secondary battery

$$U_c = U_{oc} + \eta_+ + \eta_- + \Delta E_{ohm} \qquad (1.10b)$$

The dependence of ΔE_{ohm} on current is linear (Ohm's law), the dependence of η on current is principally non-linear. A graphical presentation of U,I dependence, the so-called current-voltage curve is the important characteristics of a battery. Often, current-voltage curves are S-shaped (Fig. 1.2), which relates to the different nature of polarization at low and high current loads.

Therefore, the formal effective internal resistance of a cell $R_{eff} = -dU_d/dI_d$ is not a constant but transforms with changing current. In many cases, however, for the sake of practical convenience, this dependence is represented by a very simplified linear equation:

$$U_d = U_0 - I_d R_{app} \qquad (1.11)$$

where it is assumed that the apparent internal resistance, R_{app}, is constant. It is a rather crude approximation, (Fig. 1.2) but nevertheless is useful.

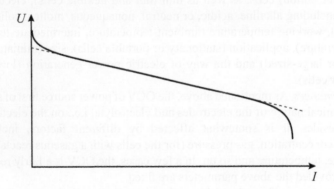

Figure 1.2. Typical current-voltage curve for discharge of a cell

In the course of discharge a cell voltage is changing, but as usual is decreasing. The plot of U_d vs the delivered charge Q_d is called the 'discharge curve', presents a very important characteristics of a battery. If the battery is discharged at constant current, the discharge curve can be presented as a plot of U_d vs time of discharge, t. Typical examples of discharge curves are shown in Fig. 1.3.

The degree of voltage fall-off is different in different cell types. The decrease of U_d on discharge can be related to decrease of OCV as well as to increase of electrode polarization and/or inner resistance. The actual shape of a discharge curve is determined by various factors. The terminal decrease of voltage at the end of discharge may be steep (Fig. 1.3a) or gradual (Fig. 1.3b). The initial phase of discharge is sometimes accompanied by some short-term increase of U_d. In other cases, U_d sharply drops and then rises again, manifesting a typical 'dip' (Fig. 1.3c).

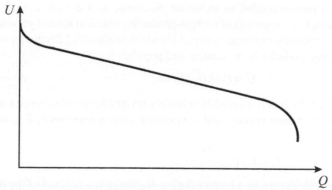

Figure 1.3a. The typical discharge curve with steep decrease of voltage at the end of discharge

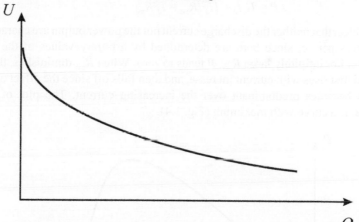

Figure 1.3b. Discharge curve with gradual change of the voltage

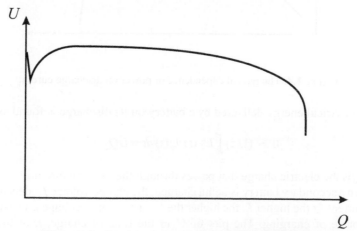

Figure 1.3c. Discharge curve with the initial voltage dip

Sometimes it is recommended to terminate discharge at a certain cut-off voltage, $(U_d)_{c-o}$, even though the reactants of current-producing reaction are not yet completely consumed. The parameter of mean integral (or mean arithmetic) discharge voltage $\bar{U}$ over the discharge period τ is convenient and popular:

$$\bar{U} = (1/\tau)\int_0^\tau U(t)dt \qquad (1.12)$$

The current, which is delivered by a battery on discharge (discharge current I_d), depends on the discharge voltage and the external circuit resistance, R_{ext}, according to Ohm's law:

$$I_d = U_d/R_{ext} \qquad (1.13)$$

The power P delivered by a battery during discharge is a product of the discharge voltage and the discharge current:

$$P = U_dI_d = U_d^2/R_{ext} = I_d^2R_{ext} \qquad (1.14)$$

It is clear that neither the discharge current nor the power output are characteristics of a battery per se, since both are determined by arbitrary values of the external resistance. For infinitely large R_{ext}, P tends to zero. When R_{ext} diminishes, the power delivered first rises with current increase, and then falls off since the effect of voltage decrease becomes predominant over the increasing current. The plot of P vs I_d, therefore, is a curve with maximum (Fig. 1.4).

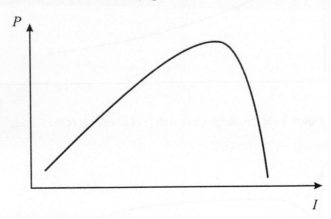

Figure 1.4. The typical dependence of power on discharge current

The electrical energy delivered by a battery on its discharge is found as

$$W = (1/\tau)\int_0^\tau U_d(t)\,I_d(t)\,dt \approx \bar{U}Q_d \qquad (1.15)$$

where Q_d is the electric charge that passes through the external circuit.

When a secondary battery is being charged, the charge voltage U_c depends on the charge current I_c: the higher I_c, the higher the U_c. The charge voltage is also increased in the course of charging. The plot of U_c vs the time of charge, t, or the electric charge, Q_c, is called the 'charge curve'. Typical examples of charge curves are shown in Fig. 1.5. A battery is charged either until a certain cut-off voltage is reached (often it is marked by some voltage hike) or until the battery accepts a prescribed charge Q^0 defined as 'rated capacity'.

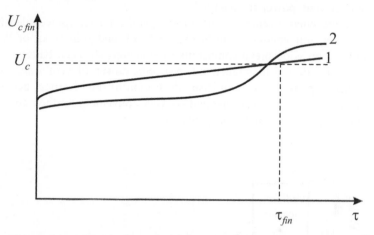

Figure 1.5. Typical charging curves without (1) and with (2) a hike at the end of charging

As a rule, the capacity of a secondary battery is referred to one complete discharge after this battery has been charged to a full rated capacity. In practice, however, it can occur that a battery is charged following an incomplete discharge. In this case, the depth of discharge (DoD.) $\theta_d \equiv Q_d/Q^0$ is less than the unity. The quantity $\theta_c = 1 - \theta_d$ is called a 'state of charging'.

In the ideal case, the discharge capacity of a secondary battery must be equal to its charge capacity. However, the charging and discharging processes is usually accompanied by various side processes. So, Q_d is less than Q_c, and the so-called Coulombic efficiency $\mu_Q = Q_d/Q_c$ is less than unity. Another important characteristic of a secondary battery is its watt-hour efficiency $\mu_w = \overline{U}_d Q_d / \overline{U}_c Q_c$. As the discharge voltage is always less than the charge voltage, μ_w is always less than μ_Q.

Usually, a battery current is expressed as the so-called C-rate. The C-rate is numerically a fraction or a multiple of the rated capacity of the battery expressed in Ah. Thus, for the battery with rated capacity 5 Ah, the C-rate is 5 A, the 3C-rate is 15 A, the C/5-rate is 1 A, and so on. Another quantity convenient for comparing different charge or discharge modes is the inverse value, specifically the time needed for a complete charge or discharge of the battery (e.g., a 2 hours discharge corresponds to C/2-rate).

Normalized performances: For comparison of electric and other characteristics of batteries differing in their size, design or electrochemical system, normalized (reduced) parameters are used. The current-producing reactions occur on the surface of electrodes in contact with the electrolyte. The current density is known to be a measure of the reaction rate. The plots of voltage against current density provide a useful characterization of a battery reflecting its specific properties independent of its size. Not only the current but also other extensive characteristics of the battery, such as power, capacity or energy, can be referred to the unit of the surface area (W/

cm², Ah/cm², or Wh/cm²). The apparent internal resistance can likewise be referred to unit of the surface area, but the relevant unit of measurements will be Ohm×cm² due to the obvious fact that with a rising surface area the internal resistance decreases in contrast to current, power and energy.

The specific energy (energy density) or specific power (power density) per unit mass or per unit volume (w in Wh/kg or Wh/L and p in W/kg or W/L) are very important and widespread comparative parameters. In each battery type, the specific energy is a diminishing function of a specific power. Plots of w against p (the so-called Ragone plots [2]) yield a clear illustration of the electrical performance parameters of given types of batteries and are very convenient for their comparison. An example of the Ragone plot is presented in Fig. 1.6.

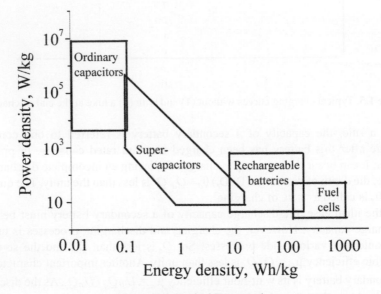

Figure 1.6. Typical Ragone plot

Generally, an energy density is independent on the battery size (and weight). However, for small-scale batteries with their inherently low absolute energies both energy density and power density are considerably reduced. This fact is explained by an increased fraction of the battery volume and weight accounted for by structural parts (the case, current-collectors and so on) and the correspondingly reduced fraction shared by the active materials.

1.1.2 Traditional secondary batteries

Lead-acid batteries. Lead-acid batteries were the first commercialized kind of rechargeable batteries. They were invented as early as 1859, but are still important up to now. Each car and truck is equipped by a SLI battery, providing not only starting, lighting and ignition but also a lot of other functions required in electrical energy. Besides SLI batteries there are also traction and stationary batteries. At present the yearly consumption of lead for manufacture of lead–acid batteries amounts to 80% of the total world output. Lead–acid battery has a good performance, high reliability and is low in price.

The electrochemical system of charged lead–acid battery is

$$(+)PbO_2|\ 30-40\%H_2SO_4(aq)\ |Pb(-)$$

The current-producing processes are as follows:

$$(+)\ PbO_2 + 3\ H^+ + HSO_4^- + 2\ e^- \underset{ch}{\overset{disch}{\rightleftarrows}} PbSO_4 + 2\ H_2O \qquad (1.16a)$$

$$(-)\ Pb + HSO_4^- \underset{ch}{\overset{disch}{\rightleftarrows}} PbSO_4 + H^+ + 2\ e^- \qquad (1.16b)$$

In a charged lead–acid battery, the active material of negative electrode consists of sponge lead and that of positive electrode contains lead dioxide. Discharge of the lead-acid battery consumes sulfuric acid and produces water and barely soluble lead sulfate on both electrodes. The concentration of sulfuric acid varies from 30–40% in a charged battery to 12–24% at the end of discharge.

Stationary lead–acid batteries possess rather low energy density (10–15 Wh/kg) but maximal service life (up to 20 years). SLI batteries have the highest energy density (25–30 Wh/kg). Their service life is as small as 2–5 years and cycle life is 200–400 cycles. Traction batteries occupy an intermediate position: 20–30 Wh/kg, 4–6 years, and 1000–1500 cycles. All the above figures refer to the rather low C-rate (ca C/5) and room temperature.

Batteries with Nickel Hydroxide Positive Electrode. There is a family of rechargeable batteries with aqueous alkaline electrolyte, the positive electrode is based on higher nickel hydroxide and various negative electrodes, namely cadmium metal, iron metal and metal hydride. The electrochemical systems of such batteries in a completely charged state are as follows:

NiOOH | 20–22% KOH (aq) | Cd (nickel–cadmium batteries)

NiOOH | 20–22% KOH (aq) | Fe (nickel–iron batteries)

NiOOH | 20–22% KOH (aq) | MeH (nickel–metal hydride batteries), where MeH stands for metal hydride.

The history of nickel-cadmium and nickel-iron batteries was started between the 19th and 20th centuries, the first reliable nickel–metal hydride battery was patented in 1975.

The main current-producing reactions on the positive electrode can be expressed as

$$(+)\ 2NiOOH + 2H_2O + 2\ e^- \underset{ch}{\overset{disch}{\rightleftarrows}} 2Ni(OH)_2 + 2\ OH^- \qquad (1.17)$$

The current-producing reactions on the negative electrodes are as follows:

$$Cd + 2\ OH^- \underset{charge}{\overset{discharge}{\rightleftarrows}} Cd(OH)_2 + 2\ e^- \qquad (1.18)$$

$$\text{Fe} + 2\,\text{OH}^- \underset{\text{charge}}{\overset{\text{discharge}}{\rightleftharpoons}} \text{Fe(OH)}_2 + 2\,\text{e}^- \qquad\qquad (1.19)$$

$$\text{MeH} + \text{OH}^- \underset{\text{charge}}{\overset{\text{discharge}}{\rightleftharpoons}} \text{Me} + \text{H}_2\text{O} + 2\,\text{e}^- \qquad\qquad (1.20)$$

The active material of negative electrodes of nickel–metal hydride batteries is a multi component alloy, which can reversibly take up considerable quantities of hydrogen into their crystal lattice, i.e., form metal hydrides. Usually, such alloys contain rare-earth metals, e.g., LaNi or $MmNi_{3.5}Co_{0.7}Mn_{0.4}Al_{0.3}$, where Mm (misch metal) is a non separated mixture of rare-earth metals.

The energy density of nickel–cadmium and nickel–iron batteries amounts to 20–40 Wh/kg, and the energy density of nickel–metal hydride batteries is notably higher and reaches up to 60–80 Wh/kg. The cycle life of nickel–cadmium batteries can be as high as 2500 cycles.

1.2. Lithium *vs.* lithium-ion batteries

A principally new type of chemical power source, namely, lithium cells with non-aqueous electrolytes were commercialized in the early 1970's and later earned very wide acceptance. Thus, a long established dream of scientists, viz., the creation of a chemical power source with the most active reductant, i.e., an alkali metal, was realized. Lithium possesses the most negative working potential and simultaneously, the highest specific capacity. The use of such a reductant allows one to enhance both the operational voltage of a power source and its energy density. In contrast to primary cells with lithium anodes, which were developed relatively quickly and successfully and are now used as the power sources in portable devices of everyday and special equipment, the history of the development of rechargeable lithium batteries has been very dramatic.

There are two kinds of primary cells designed for operation at ambient temperatures, namely cells with a liquid cathode (oxidizer) and cells with a solid cathode. Thionyl chloride and sulfur dioxide, which are also the basis of electrolytes, are most often used as liquid oxidizers, that are simultaneously utllized by solvents of the electrolytes. Such cells have relatively high specific power values, they have the highest energy density, are operable in a wide range of temperatures, but they are fire and explosion hazards and are used only in products of special (military) equipment. In cells with solid cathodes, oxides of manganese, copper or some other metals, as well as ferrous sulfide and fluorinated carbon are used as cathode active materials. Solutions of some lithium salts (perchlorate, hexafluoroarsenate, tetrafluoroborate, etc.,) in organic non-aqueous (aprotic) solvents (propylene carbonate, dimethoxyethane, tetrahydrofuran, gamma-butyrolactone and various mixtures) are used as electrolyte in these cells. Depending on the type of cathode material, the operating voltage of such cells is close to 3 V (when using manganese dioxide or fluorinated carbon) or 1.5 V (when using copper oxide or iron sulfide); in the latter case, the lithium cells are interchangeable with the traditional cells of the manganese-zinc system. It is the primary cells with aprotic electrolyte that appear to be the closest analogs of rechargeable (secondary) cells.

The chemistry of secondary cells with aprotic electrolytes is very close to the chemistry of primary cells. The same processes of anodic lithium dissolution at the negative electrode and cathode insertion of lithium into the crystal lattice of the material of the positive electrode occur at the discharge of such cells. When charging the secondary cells, the electrode processes should proceed in the opposite direction. At the end of the 70's, materials of a positive electrode were found, on which the cathodic insertion and anodic extraction (i.e., cathode intercalation and anodic deintercalation) of lithium proceed almost reversibly. Examples of such compounds are titanium disulfide or molybdenum disulfide.

The main problem appears with a negative electrode. When it is charged, i.e., during cathode deposition of lithium, complications typical of galvanic practice appear. The surface of lithium in aprotic electrolytes is known to be covered with a passive film due to the chemical interaction with electrolyte components, specifically an organic solvent and anions. This film has the properties of a solid electrolyte with lithium ion conductivity. The film is quite thin (its thickness does not exceed units of nanometers) and can well protect lithium from self-discharge, i.e., from interaction with an electrolyte. During cathode deposition of lithium, a fresh, very active surface is formed, on which a passive film grows, and since lithium is deposited in the form of dendrites, in many cases in charge-discharge cycles, the film completely envelops individual lithium microparticles, preventing their electronic contact with the current collector. This phenomenon is called 'encapsulation'. Encapsulation leads to the fact that with each cycle a certain part of the lithium is eliminated from further work. Therefore, an excess in comparison with the stoichiometric amount of lithium must be added to the secondary cells with a metallic lithium electrode. This excess is from 4 to 10-fold, thus, the effective specific capacity of lithium decreases from the theoretical value of 3828 mAh/g to the values of 380–800 mAh/g. In addition, dendritic formation leads to the danger of short circuits, i.e., to fire and explosion hazards of such devices. A lot of effort has been directed to find different methods of surface treatment (or the introduction of appropriate additives into the electrolyte), which could prevent dendrite formation during cathodic deposition of lithium. It was found, for example, that the addition of small (trace) amounts of hydrogen fluoride to the propylene carbonate electrolyte contributes to the deposition of lithium not in the form of dendrites, but in the form of a dense, fine-grained solid layer.

Over time the beneficial effect of lithium surface treatment with carbon dioxide (or simply introducing CO_2 itself or products emitting it when interacting with lithium into the electrolyte) have been pointed out. It was also proposed to modify the lithium surface by applying monomolecular layers of non-ionic surfactants substances, in particular, dimethyl ether of polyethylene glycol or a copolymer of dimethylsiloxane with propylene oxide. In addition, it was found that in some specially selected electrolytes dendrite formation does not occur even in the absence of additives. An example of such an electrolyte is a solution of lithium hexafluoroarsenate in 1,3-dioxolane with tributylamine addition.

Despite the fundamental difficulties of creating batteries with a metal lithium electrode, there have been attempts to commercialize such products. Thus, in 1987, the Moli Energy Ltd., Canada began a serial production of AA-size batteries using positive electrodes of molybdenum disulfide. The batteries had an initial voltage of

2.3 V, and a capacity of 600 mAh at a discharge current of 0.2 A to a final voltage of 1.3 V. In some cases, when discharging to a final voltage of 1.1 V, it was possible to get a discharge in the first cycle capacity up to 800 mAh. According to the company, batteries had a cycle-life of several hundred cycles, however, these data were disputed by a number of experts and consumers. Actually, it was possible to talk about a cycle-life of 100 cycles. In 1989, an advanced battery was created using molybdenum sulfide of a different composition (Mo_6S_8). These batteries had a higher discharge voltage (from 2.3 to 1.8 V) and higher capacity (up to 1000 mAh), and also allowed forced discharge with a current of up to 1 A.

The characteristics of Moli's batteries might have been acceptable, despite the large voltage drop during discharge and low actual life, but the reliability of the batteries was insufficient. In the summer of 1989, isolated cases of ignition of primary lithium cells were recorded, as well as Moli secondary batteries failed and the production of such batteries was discontinued.

At the Jet Propulsion Laboratory of the California Institute of Technology (USA), secondary cells with positive electrodes of titanium disulfide have been developed but they did not go for mass production, mainly because of the relatively small cycle life (100-200 charge-discharge cycles). In 1989 Grace & Company (USA) claimed on the fabrication of AA-size spirally bound cells with titanium disulfide. Such cells had a cycle life from 50 to 200 cycles. AT&T Bell Laboratories (USA), reported on development of AA-size cells with niobium triselenide as an active material of positive electrode. The reported cycle life was about 250 cycles. Approximately the same situation appeared with the development of batteries at Honeywell and Eveready Battery Co. Inc. In the Matsushita Electric Industrial Co. Japan, AA-size batteries were developed using manganese dioxide as the active material of the positive electrode. The batteries had a discharge voltage of 3.2 to 2.0 V and a capacity of 800 mAh at the beginning of operation, 600 mAh after 60 cycles, and 450 mAh after 200 cycles. All these attempts have now only a historical interest.

In attempts to overcome the problems associated with the use of lithium metal in negative electrodes, some lithium alloys were tested. Li-Al alloys were presumed to be the most promising candidates. During the discharge of such an electrode, the lithium concentration decreases (lithium is etched out of alloys) and during the charge it increases again. The chemical activity of lithium in an alloy is somewhat lower than that of the lithium metal, and the potential of an alloy-based electrode is more positive than the potential of pure lithium. The consequences are on the one hand, a decrease in the working voltage and, on the other hand, a weakening of the alloy interaction with the solvent, i.e., reducing the self-discharge.

Matsushita Electric Industrial Co. (Japan) developed disk-shaped rechargeable batteries of the VL 2020 with negative electrodes from a Li-Al alloy and positive electrodes based on vanadium oxide. Such batteries demonstrated the cycle life of about 1000 charge-discharge cycles for a 10% Depth of Discharge (DoD) and about 450 cycles for a 20% DoD. At the Jet Propulsion Laboratory of the California Institute of Technology (USA), similar cells were built with advanced characteristics: cycle life of about 600 cycles at 50% DoD. Unfortunately, Li-Al alloys suffer from substantial changes of their specific volumes during the cycling. A deep discharge results in embrittlement and crumbling of the electrode. That is why, no further progress was observed in the field using Li-Al alloys as the material for negative electrodes.

The announcement on the launching production of lithium-ion batteries with negative electrodes of carbon materials marked a revolution in the field of rechargeable lithium batteries. Carbon was shown to be a very convenient matrix for lithium intercalation. The specific volumes of graphitized materials change by no more than 10% on intercalation of sufficiently large amounts of lithium.

The potential of carbon electrodes containing moderate amounts of intercalated lithium can be 0.3-0.4 V more positive than that of the lithium metal potential. To make the cell's voltage sufficiently high, the inventors of the first lithium-ion batteries used lithiated cobalt oxide as the active material for positive electrodes. The potential of lithiated cobalt oxide was marginally higher than 4 V vs. lithium electrode, therefore the working voltage of the first lithium-ion batteries was about 3.6-3.7 V. With the battery charging, extraction of lithium ions from lithiated oxide on the positive electrode appears. These ions are transferred towards the negative electrode, where their intercalation into carbon takes place. During discharging the processes are reversed. In which case, there is no lithium metal in the battery, and the discharge and charging processes are reduced to shuttling of lithium ions from one electrode to another. That is why such batteries were named 'rocking-chair' or 'lithium-ion' batteries. Lithium-ion batteries proved to be very successful. Their development was very fast, and now they have replaced all other types of secondary batteries in portable equipment, mainly in the so-called 3C family (computations, communication and cameras). The modern annual production of lithium-ion batteries is measured by billions of pieces and by dozens of billions of dollars. The present-day application areas for lithium ion batteries is far wider than the original small electronic devices for the 3C market. Firstly, the nomenclature of small-size devices was enlarged (toys, e-cigarettes and vaporizers, lighting (LCD and fluorescent lights), medical devices, garden tools, and so on). Secondly, the main problem is development of battery-powered electric vehicles.

The negative electrode in all modern commercial lithium-ion batteries is manufactured from special carbon materials. Under cathodic polarization lithium ions intercalate into carbon crystal lattice with charge transfer. Simultaneously, the electrolyte (both the solvent and various impurities) is reduced. The products of this reduction are insoluble in the electrolyte and form a passive film on the carbon surface. It is very important that this passive film possesses the property of a solid electrolyte. Therefore, charge in the first cathodic run is consumed both in intercalation of lithium and formation of a passive film. After this film is fully formed, it prevents the direct contact between the electrolyte and carbon and terminates further electrolyte reduction.

In the case of graphite, the thermodynamically stable LiC_6 compound is obtained at lithium intercalation. The equation of intercalation–deintercalation is usually as follows:

$$x \, Li^+ + x \, e + 6 \, C \leftrightarrow Li_xC_6 \qquad (1.21)$$

According to Faraday's laws the specific capacity of graphite amounts to 372 mAh/g. Lithium is intercalated into graphite below rather negative potentials (more negative than 0.5 V *vs.* Li^+/Li). Therefore, the activity of lithium in such intercalates is rather high.

Over three decades that have passed since the commercialization of lithium-ion batteries, repeated attempts have been made to improve the negative electrode. One of the most successful materials proposed for this purpose was silicon. (See Chapter 2 for more detailed information on using silicon in all-solid-state batteries). Silicon is very suitable for thin-film batteries, and in traditional batteries it is used as an additive to carbon. The composition of carbon materials with minor nano silicon additives provides specific capacities from 400 to 500 mAh/g.

The positive electrodes of lithium–ion batteries are based on lithiated cobalt or nickel oxides and of lithium–manganese spinels.

The functioning of the positive electrode consists in lithium extraction under battery charging (when positive electrode is oxidized) with a corresponding increase in the metal valency and to insertion of lithium under discharge:

$$LiCoO_2 \leftrightarrow x\ Li^+ + x\ e + Li_{1-x}CoO_2 \qquad (1.22)$$

When extracting more than 50% lithium from $LiCoO_2$, irreversible structural changes occur. Therefore, electrodes based on pure $LiCoO_2$ are not cycled to a full depth, but only within $0 < x < 0.5$, which corresponds to a specific capacity of about 140 mAh/g. Significant drawbacks of $LiCoO_2$ are its rather high cost and toxicity. In recent years, multi component oxides ($LiNi_xCo_{1-x}O_2$, $LiNi_{1-x-y}Co_xAl_yO_2$, $LiNi_{1-x-y}Mn_xCo_yO_2$) were used more often as cathode material, of which $LiNi_{1-x-y}Mn_xCo_yO_2$ (NMC) is the most popular. Electrodes with such material have a specific capacity of upto 240 mAh/g. A highly competitive material also is $LiNi_{0.80}Co_{0.15}Al_{0.05}O_2$(NCA).

Lithium-manganese spinel $LiMn_2O_4$ are most often considered among lithiated manganese oxides. At charge, i.e., under lithium extraction, $Li_{1-x}Mn_2O_4$ type compounds are formed. Their working potential is close to 4 V.

A significant alternative to oxide cathode materials is lithium iron phosphate $LiFePO_4$. At the moment about 25% of lithium-ion batteries are based on this very material. During anodic oxidation (charge) of $LiFePO_4$, iron phosphate $FePO_4$, which is isostructural to the initial $LiFePO_4$ is formed:

$$LiFePO_4 \leftrightarrow FePO_4 + Li^+ + e \qquad (1.23)$$

Due to structure closeness of $LiFePO_4$ and $FePO_4$ no significant structural changes on cycling of lithium iron phosphate electrodes occur. Such a feature of $LiFePO_4/FePO_4$ system assures a very good cyclability of iron phosphate electrodes (several thousands of full cycles). During charge or discharge no compounds with variable composition are formed, that is why the discharge and charge occur at a practically invariable potential.

$LiFePO_4$ possesses very low electron conductivity (less than 10^{-9} S/cm). It is for this reason that the electrode material is manufactured in the form of 20–30 nm particles. In addition, the surface of these particles is completely covered by a carbon layer with the thickness of several nm. The equilibrium potential of the $LiFePO_4/FePO_4$ system is close to 3.5 V vs. the lithium electrode, i.e., it is considerably less positive than in the case of oxide cathode materials. This eliminates or simplifies the problem of electrolyte stability and enhances the safety of batteries.

In most lithium-ion batteries 1.0–1.3 M $LiPF_6$ in mixed solvent containing ethylene carbonate, dimethyl carbonate and diethyl carbonate is used as an electrolyte.

1.3. The features of thin-film batteries

The concept of thin-film batteries is closely related to the concept of all-solid-state batteries. The need for such devices appears from the rapidly developing microelectronics, especially battery-powered Integrated Circuit (IC) cards (smart-cards), radio-frequency identifier (RFID) tags, smart watches, implantable medical devices, remote microsensors and transmitters, Internet of Things (IoT) systems, and various other wireless devices including smart building control and so on. Some important kind of thin-film batteries are flexible ones. By definition, the thickness of thin-film batteries is much less than their length and width. If the thickness of such batteries is usually measured in tens of microns, then their area may be as small as is achievable with modern lithographic techniques, and as large as square meters. Such a large range of the batteries' size provides a wide range in their capacity, and a variety of their applications.

The absence of a liquid electrolyte is an important advantage of thin-film all-solid-state batteries, since it increases their intrinsic safety. The absence of organic solvents eliminates the risk of possible fluid and vapor leakage and therefore reduces the risk of fire and explosions. Moreover, the solvents of the liquid electrolytes often participate in the degradation processes of traditional lithium-ion batteries, that is why the cycle-life performance of solid-state batteries is supposed to be much larger. The thickness of the solid electrolyte as a rule is less than the thickness of the separator for liquid electrolytes. At the same time, the small thickness of the solid electrolyte is also dictated by its relatively high resistivity. Typical thickness of common separators in lithium-ion batteries is about 20 μm, whereas the thickness of solid electrolytes amounts to 1 μm.

The schematic of thin-film battery is depicted in Fig. 1.7.

The insignificant thickness of a thin-film battery forces it to be placed on a more or less large structural element (substrate), which can be part of a device powered by this battery. And it is, perhaps, the main difference between thin-film batteries and usual commercial lithium-ion batteries. The second principal difference consists in the possibility of using lithium metal as a negative electrode in all-solid-state batteries. It is known that the key feature of lithium-ion batteries consists in using intercalation electrodes instead of lithium metal.

The battery's substrate, in principle, can consist of any material, including metals, ceramics, glasses, polymers and even paper. If this material is an electronic conductor, the substrate can play the role of a current-collector of one electrode (usually, lithium). In any case, the material of the substrate must meet the conditions at which the functional layers are deposited and operated. The substrate material must not reveal any interaction with the other layers of the battery. The substrate material must also hinder a lithium diffusion out of the battery. The battery per se consists of two electrodes with an electrolyte sandwiched in between. The outer side of each electrode contacts with the corresponding current collector. The battery as a whole is enclosed in an appropriate packaging structure (case). The case has a

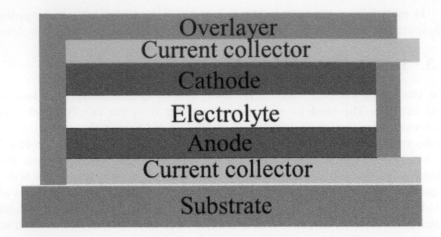

Figure 1.7. The principal design of thin-film all-solid-state battery

rather important design element. It must provide protection of the internal contents of the battery against external physical and chemical influences, in particular, it must prevent harmful reactions of the battery materials with air and moisture. Ideally the battery and its powered electronic device should be functionally integrated with maximum efficiency and voltage control.

The interface between an electrolyte and positive electrode in Fig. 1.7 is the plane surface, whose area is rather limited, that corresponds to rather high current density. It is quite desirable to reduce the true current density, and therefore to lower polarization losses. This can be achieved by making use of the enlarged surface area of 3D micro batteries.

These days there are a number of excellent review papers devoted to all-solid-state lithium-ion batteries [3–8].

The first all-solid-state batteries systems were developed during the 1950's. They did not use strong reducers and oxidizers, their electrolytes had rather low conductivity, and the batteries had low voltages which limited their applicability drastically. Primary solid-state cells with a lithium anode appeared almost simultaneously with their liquid-electrolyte-counterparts in the 60's of the last century. These cells were based on lithium iodide as a solid electrolyte [9,10]. The first cell was based on the electrochemical system

Li / LiI / AgI

and was made by the following steps: (*i*) AgI was sprayed on a silver foil and then AgI was melted; (*ii*) LiI electrolyte was vacuum deposited onto the AgI film; (*iii*) Li was vacuum deposited onto the electrolyte. The thicknesses of the components of the cells were the following: silver foil, 13 µm; AgI cathode, 30 µm; LiI electrolyte, 15 µm; lithium anode, 5 µm. An open-circuit voltage of the cell was 2.1 V. Its polarization was purely Ohmic and corresponded to specific conductivity of LiI ($8.5*10^{-8}$ S/cm).

Soon after that the electrochemical system Li / LiI / CuI was changed to Li / LiI / AgI. In this case the cathode was made by pressing a mixture of CuI and LiI powders on a thin copper foil. Such a mixture of active material (CuI) and electrolyte (LiI) in the cathode active layer manifests a principle of 3D design.

Some improvement of cells based on electrochemical system Li / LiI / AgI was achieved by changing LiI by $4LiI \cdot NH_4I$, and replacement of AgI by $NH_4Ag_4I_5$. Subsequently, it was proposed to dope LiI with various additives, which led to a significant (by two orders of the magnitude) increase in the conductivity of the solid electrolyte.

The next step in improving the LiI electrolyte was its doping with CaI_2. This approach was based on the fact that doping by divalent cations induces some defects in the crystal lattice of lithium halogenides, with a corresponding increase of ion conductivity. Unfortunately, the solubility of CaI_2 in LiI is rather limited, so such a doping resulted in ionic conductivity increase by one order of magnitude only. Later the nomenclature of dopants was enlarged, and specific conductivity of doped LiI electrolyte was increased up to 10^{-5} S/cm.

An interesting idea was using elemental iodine as a cathodic material instead of silver iodide. The serious drawback of elemental iodine consists in its relatively high vapor pressure. Besides, iodine can travel through the electrolyte towards the anode causing self-discharge. That is why some charge transfer complexes were proposed instead of iodine. Such a complex contains an organic molecule as an electron donor, and a halogen as an electron acceptor. Various organic compounds were considered as ligands, and at this point the most popular ligand is poly-2-vinylpyridine $[CH_2CH(C_5H_5N)]_n$ (P2VP). The composition of the complex can be expressed as $P2VP \cdot mI_2$. During cell discharge the iodine content in the complex decreases and solid lithium iodide is formed:

$$P2VP \cdot mI_2 + x \, Li^+ + x \, e \rightarrow P2VP \cdot (m-x/2)I_2 + x \, LiI \qquad (1.24)$$

In 1983 a new thin-film all-solid-state battery with a glass-like electrolyte and TiS_2 as cathode material was described (Hitachi Ltd., Japan) [11]. The substrate of the battery was made from a silica glass. A layer of TiS_2 was fabricated on this substrate by low pressure Chemical Vapor Deposition (CVD) by using $TiCl_4$ and H_2S as a source of gases. Then thin-film amorphous electrolyte with composition $Li_{3.6}Si_{0.6}P_{0.4}O_4$ was applied onto the TiS_2 surface by radio-frequency magnetron sputtering at a pressure of 3 Pa, by using $Li_{3.6}Si_{0.6}P_{0.4}O_4$ powder and Li_2O pellets as a target, and a mixture of argon and oxygen (3:2) as a sputtering gas. Finally, lithium film was vacuum evaporated at a pressure of 10 Pa. The ionic conductivity of the glass electrolyte was $5*10^{-6}$ S/cm at room temperature. At the end of the 1980's another glass-like electrolyte was proposed, namely $Li_{3.4}V_{0.6}Si_{0.4}O_4$ (Nippon Telegraph and Telephone Corp., Japan) [12]. Sulfide glass-like electrolytes with a composition, e.g., $Li_3PO_4–P_2S_5$, $6LiI–4Li_3PO_4–P_2S_5$ were proposed as well (Eveready Battery Co., USA) [13]. An alternative version of glass-like electrolyte was lithium borosilicate electrolyte, based on a system $Li_2O-B_2O_3-SiO_2$ [14]. But the most popular glass-like electrolyte was lithium phosphorus oxynitride (LiPON) [15–18]. The LiPON is prepared by magnetron radio-frequency sputtering of Li_3PO_4 target in nitrogen gas. Its average composition could be expressed as $Li_{3.3}PO_{3.8}N_{0.22}$, with some uncertainty of the nitrogen content. The addition of nitrogen to the glass structure was supposed to enhance the chemical and thermal stability of a lithium glass. LiPON is rather stable in contact with lithium metal, possesses very low electronic conductivity and adequate ionic conductivity of $2.3*10^{-6}$ S/cm at room temperature with lithium transport number equal to unity. The decomposition voltage of LiPON was shown to be higher than 5.5 V. The team from Oak Ridge National Laboratory (ORNL, USA) achieved the most success with such an electrolyte. Using this very electrolyte, thin-film rechargeable all-solid-state lithium batteries were fabricated with a variety of active cathode materials including $Li_xMn_2O_4$, TiS_2, $LiCoO_2$, and V_2O_5. V_2O_5 enabled to achieve the most impressive results at that time.

The substrates of these batteries were made usually from glass (ordinary microscope slides), but plates of alumina and 0.1 mm-thick polyester were used as well. Anode and cathode current collectors from vanadium metal were deposited by dc-magnetron sputtering in argon. Then active materials were deposited. Vanadium pentoxide was applied by dc-magnetron sputtering by using vanadium target and a mixture of argon with 14% oxygen as a working gas. The cathode thickness was about 1 μm. The layer of LiPON electrolyte with thickness of 1 μm was applied to the cathode by rf-magnetron sputtering with a target from Li_3PO_4 and nitrogen as a working gas. The lithium anodes were deposited to the electrolyte by evaporation of lithium metal in vacuum from a Ti crucible. The Li films were typically 3-5 μm thick, therefore it was notably stoichiometric excess of lithium in comparison with vanadium pentoxide. Total thickness of both electrodes and electrolyte amounts to 10–15 μm, that is comparable with the support thickness

The battery demonstrates reasonable energy densities, the ability to undergo thousands of charge-discharge cycles. The temperature range of the batteries' operation is limited by lithium melting point only (180 °C). The reasonable working current density of such batteries is modified to about 0.1 mA/cm², and this limitation is dictated by the slowness of Li^+ insertion into the cathode material.

By the middle of the 2000's several companies had commercialized such -solid-state batteries with a capacity from 0.1 to 5 mAh. STMicroelectronics (France), Geomatic Co. Ltd. (USA), Excellatron (USA), Cymbet Corp. (USA) that need to be mentioned. The batteries were able to achieve various applications, including smart cards, sensors, RFID tags, microelectromechanical systems (MEMS), and so on. The technique of thin layer deposition was not limited by thermal evaporation and dc or rf magnetron sputtering. Pulsed laser deposition, electron cyclotron resonance sputtering, aerosol spray coating, electrostatic spray deposition, ink-jet printing, electrophoretic deposition, etc., were used with more or less success.

In most cases lithium metal anode is used in thin-film batteries. It is very important that such anodes are in contact with a solid electrolyte demonstrating excellent charge–discharge cycling at high rates, and a long battery life with no lithium dendrites. An interesting version is the so-called 'lithium-free' batteries. In this case the battery is assembled without lithium, but with some excess of lithiated material of cathode (e.g., $LiCoO_2$). During the first charge lithium is transferred from a positive electrode through an electrolyte and deposited to a current collector of negative electrode.

Besides lithium metals usually lithium-ion anodes that have been proposed including thin films of Sn, Si, Ge and C. It is well-known that very thin films (*ca.*50 nm), can tolerate extreme volume changes at insertion and extraction of large amounts of lithium. A good performance was also reported for various oxide, nitride and oxynitrides. A popular compound such as, $Li_4Ti_5O_{12}$, also could be used for thin film lithium-ion batteries.

As to active materials of positive electrodes (cathodes), amorphous or nanocrystalline $LiCoO_2$, V_2O_5,$LiMn_2O_4$ and $LiFePO_4$ were proposed.

Thin-film batteries with LiPON electrolyte have been cycled to hundreds and even thousands of cycles with low degradation. Such excellent cyclability could be attributed to a combination of several factors. First, the high stability of the LiPON electrolyte film. Second, the ability of the thin film materials to accommodate the volume changes associated with lithiation and delithiation. Third, the uniformity of the current distribution in the thin film structure.

An important advantage of thin-film batteries is their ability to operate in a wider temperature range than usual lithium-ion with liquid electrolytes. Good cycling performances were reported for such external temperatures as –40 and +150 °C. The decrease in temperature, of course, leads to a slowdown in all activation processes of mass- and charge-transfer. However, no physical damage of the battery due to freezing occurs in this. At the same time, at high temperatures, battery degradation increases due to gradual microstructure changes in the electrodes.

At low current rates (i.e., at low power density) the energy of thin-film batteries is roughly proportional to thickness of the cathode layer (providing the battery is cathode-limited). For instance, the energy of batteries of Li/LiPON/$LiMn_2O_4$ with a thickness of cathode layer 0.3, 1 and 2 µm amounts to 70, 200 and 400 µWh/cm² at power density 0.1 mW/cm². The thinner the cathode layer, the higher power density can be achieved.

A typical example of electrical characterization of all-solid-state thin film batteries of the 2000's can be found in [19]. The commercial battery with dimensions 20 x 15 x 0.015 mm and rated capacity 0.4 mAh, having the electrochemical system Li/LiPON/$LiCoO_2$ was tested. The schematic of the battery is shown in Fig. 1.8.

At room temperature the battery withstood 100 full-depth charge-discharge cycles with C/4 rate and cut-off voltages 4.2 and 3.0 V. Average real capacity at ever cycle was 404 µAh, and capacity loss on cycling was below the detection limit. The coulombic efficiency was strictly 100%. At room temperature the battery demonstrated excellent rate capability: discharge capacity on cycling at 2.5C was as high as 90% from that at C/4. Lowering of temperature resulted in decrease of discharge capacity, but even at –50 °C the capacity was measurable amounting to 0.013 µAh.

An unusual method of manufacturing all-solid-state battery is described in [20]. Here a layer of Li_2O–Al_2O_3–TiO_2–P_2O_5-based glass–ceramic solid electrolyte was sandwiched in between copper and manganese electrodes. To this end a copper film with a thickness of 0.5 µm was deposited on one side of the solid electrolyte sheet by radio frequency magnetron sputtering, and after that a manganese film with thickness of 1.5 µm was deposited on the other side of the electrolyte sheet by vacuum evaporation. Then direct current with voltage 16 V was applied to the

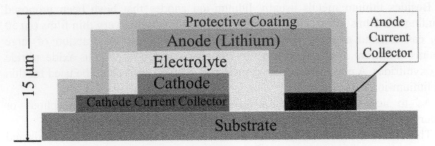

Figure 1.8. The schematic of thin-film micro battery

resultant Cu/solid electrolyte/Mn system, the Cu side being connected to the cathode, and the Mn side to the anode. With the result of electrochemical processes, a layer of manganese dioxide was formed in the interface between the manganese metal and solid electrolyte, and a certain amount of lithium metal was deposited at the surface of copper. Thus, the rechargeable battery of electrochemical system Li/MnO_2 was manufactured. (See also, the so-called 'Li-free construction', later).

For the last two decades some companies terminated production of all-solid-state batteries, whereas some others continued and developed the production. For instance, Excellatron Solid State LLC announced the following performances of their thin-film batteries: (i) Single cell discharge voltage 4 V, (2) continuous discharge currents up to 60C at room temperature, (iii) pulse discharge rate up to 1,000C at elevated temperature, (iv) cycle life 40,000 full depth-of-discharge cycles, (v) operating temperature range from –25°C to +80°C, storage temperature range from –55°C to +300°C. Such batteries are used in smart cards, active RFID tags, implantable medical devices (cardiac pacemakers, defibrillators, cochlear implants, neurostimulators, etc.,), non-volatile static random-access memory (SRAM), MEMS, and so on.

Cymbet Corporation focuses its activity on the development and large-scale production of miniature and micro miniature solid state batteries bare die to be co-packaged with other integrated circuits using wire bond attachment in a single package or multi-chip module. The company produces all-solid-state batteries with nominal capacity 50 µAh (sizing 5.7 × 6.1 × 0.2 mm), and 5 µAh (sizing 1.75 × 2.15 × 0.2 mm). The company guarantees a battery cycle-life of 500 cycles. Figure 1.9 shows typical discharge curves for a battery with nominal capacity 5 µAh at different current rates.

Data for Cymbet micro miniature batteries clearly shows how a reduction in battery size is accompanied by a significant reduction in energy density. For the two batteries mentioned above, the energy density amounts to 25 and 27 Wh/l, while for lithium-ion batteries of normal capacity (Ah units), this value is 200–400 Wh/l.

The aim of the above-mentioned 3D design is increasing the volumetric energy density of the batteries. A planar solid-state design depicted in Fig. 1.7 shows that relatively much space is occupied by inactive materials as substrate and packaging. 3D design allows obtaining much higher energy storage capacities with almost the same amount of packaging and substrate material. The internal surface area between

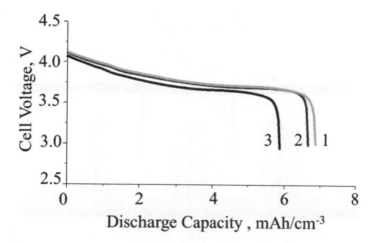

Figure 1.9. Typical discharge curves of the battery CBC005 EnerChip at the rates C/125 (1), C/12.5 (2) and 1 C (3)

electrodes and the electrolyte is enlarged in 3D batteries, therefore a much higher total battery current can be obtained with similar internal current densities. This corresponds to relatively high current- and power capability for 3D all-solid-state batteries. In fact, the energy demands of micro- and nano-electromechanical systems, including the above mentioned implantable medical devices, drug delivery systems, micro sensors and so on opened a special niche for 3D batteries with a typical size from 1 to 10 mm^3 and power from 10 nW to 1 mW. It is worth noting that there are several comprehensive reviews on 3D lithium-ion batteries [21–28].

There is a variety of concepts for a 3D micro battery design with regular or chaotic geometry. It may be a periodic array or a periodic ensemble of electrodes. For instance, it may be an array of cylindrical (columnar) electrodes of both signs grown on substrates. Two arrays of different electrodes are inter-digitated. The space between electrodes must be filled by an electrolyte (Fig. 1.10). The main disadvantage of such a design is a rather large volume of the electrolyte, and high and variable interelectrode distance.

The design consisting of an array of columnar electrodes of one sign (e. g., anodes) placed on a substrate and coated with thin layer of an electrolyte could be more effective. The remaining space in this case is filled by cathodic material.

The best method of such structure manufacturing is the use of appropriate template and usual solid-state deposition techniques. After removing the template, a 3D morphology is obtained. The template can be made from a polymer film with regular pores or anodic alumina films can play a role of the template.

Yet another design of 3D battery is described in [21]. Here a row of grooves is made in a massive silicon substrate by anisotropic etching. The substrate itself plays the role of one current collector. Subsequently, the active battery layers are coherently deposited inside this highly structured substrate, starting with an effective barrier layer, preferably TiN or TaN, to protect the substrate from lithium penetration, followed by a ~50 nm Si thin film anode, a solid-state LiPON-like electrolyte, and

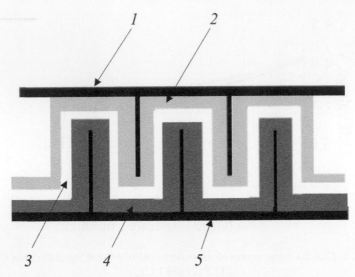

Figure 1.10. 3D design with interdigitated arrays of different electrodes.1– negative current collector, 2 – negative electrode, 3 – electrolyte, 4 – positive electrode, 5 – positive current collector

a thin film, cathode material, in this example, 1 µm LiCoO$_2$. Deposition of a second current collector completes the 3D integrated battery.

The chaotic counterpart of the design depicted in Fig. 1.10 is a kind of 'sponge' design. In this case solid network of the sponge ('web') being the cathode, is coated with very thin layer of a solid electrolyte. The remaining voids are filled with anode material.

Yet another interesting 3D design is based on micro channel plates. Here the substrate is a flat plate perforated with cylindrical holes with a radius from 15 to 50 µm. The total number of the holes may be from 8 to 30 thousand per cm^2. It is the inner surface of the holes that is the working surface of the battery, and its area is 20–25 fold the outer surface area of the plate. The substrate can be made from electron-conducting or insulating material (metals, glasses, polymers). In the latter case a current collector is deposited to the overall surface of the substrate, including inner surface of the holes and outer surface of the plate. An active cathode material is deposited onto this current collector, then a layer of an electrolyte, next an active anode material, and the finally current collector of the negative electrode. All the inner volume of the holes are filled with functional materials. This structure is mechanically more robust, than structures with free standing columnar elements. 3D design based on micro channel plates was proposed and actively developed in Tel-Aviv University [29, 30].The schematic of such battery is shown in Fig. 1.11.

The honeycomb structure presents an interesting example of inter digitated electrode configuration for 3D battery. The idea of the honeycomb structure is clear from Fig. 1.12.

Ref. [31] describes 3D micro battery with honeycomb-like LLT (Li$_{0.35}$La$_{0.55}$TiO$_3$) solid electrolyte. LLT is an example of crystal solid electrolyte. The cells in the electrolyte had dimensions 180x180x180 µm, and the gap between cells (the electrolyte

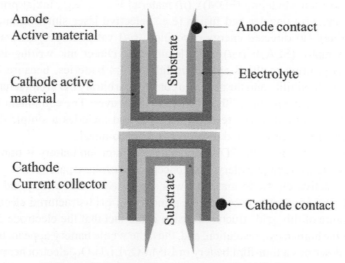

Figure 1.11. 3D design based on microchannel plate

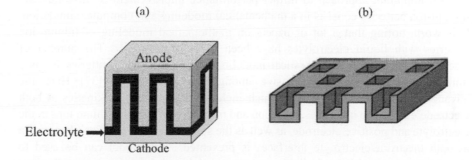

Figure 1.12. The principle of honeycomb structure (a) and view of honeycomb-like solid electrolyte (b)

thickness) was 80 μm. The described battery had an unusual electrochemical system $LiCoO_2/Li_4Mn_5O_{12}$. The discharge potential of the anode was close to 2.9 V (Li^+/Li), so the discharge voltage of the battery was about 1 V only. The active matter of the cathode and anode was placed into the alternating top and bottom cells.

Probably the most advanced manufacturing technology for 3D battery is 3D printing also known as additive manufacturing (AM) [32–34]. This technology enables full freedom of fabricating objects with well controlled and very complex geometry (e.g., thickness and shape) through a layer-by-layer deposition process directly from a Computer Aided Design (CAD) model with no need for any templates. Recently 3D printing of lithium-ion batteries with a variety of geometries has been explored and developed in order to enhance their energy density, power density and mechanical properties. In fact, 3D printing is not a single technique, but a group of techniques including: (*i*) material extrusion (e.g., Direct Ink Writing (DIW) and

Fused Deposition Modeling (FDM)); (*ii*) material jetting (e.g., inkjet printing); (*iii*) binder jetting; (*iv*) powder bed fusion (e.g., selective laser sintering and selective laser melting); (*v*) directed energy deposition; (*vi*) vat photopolymerization; (e.g., stereolithography (SLA)); (*vii*) sheet lamination. Direct ink writing is the most popular 3D printing methods applied in lithium-ion batteries, because of the low cost, material flexibility and the ability to construct arbitrary 3D architectures without additional expensive tooling, lithographic masks or dyes. The equipment for DIW is not sophisticated (and therefore not expensive) and includes a simple desktop 3D printer, heated bed, air-powered dispenser and micro-nozzle.

A notable version of the 3D all-solid-state lithium-ion battery is transparent (or semi-transparent) flexible battery. The concept of semi-transparent battery with non-transparent active electrode materials was suggested in 2011 [35] and developed later [36]. The concept is based on the principle of grid-structured electrodes. The salient feature of this grid-structured design is the fact that the electrode dimensions are below the human eye resolution, and, thus, the whole battery appears transparent. Ref. [35] describes a thin-film battery of $LiMn_2O_4/Li_4Ti_5O_{12}$ electrochemical system with gel-polymer electrolyte, whereas Ref. [36] describes a battery of $LiCoO_2/Si$ system with LiPON electrolyte. The transparency of the both batteries reaches 60%.

An important approach to further performance improvement of all-solid-state lithium-ion batteries consists in a mathematical modeling and computer simulation. It is worth noting that a lot of papers on mathematical modeling of lithium-ion batteries with liquid electrolytes have been published recently. The number of researches devoted to mathematical modeling of all-solid-state batteries is quite limited, and the most comprehensive among them is the paper [37]. Here, the advanced one-dimensional model which includes charge-transfer kinetics at both electrode/electrolyte interfaces, diffusion and migration of mobile lithium ions in the electrolyte and positive electrode, as well as the space-charge separation phenomena at both electrode/electrolyte interfaces is presented. This model can be used to simultaneously study the individual overpotential and impedance contributions together with concentration gradients and electric fields across the entire battery stack. Both galvanostatic discharge and impedance simulations have been experimentally validated with respect to 0.7 mAh $Li/LiPON/LiCoO_2$ thin film, all-solid-state battery. The model is in good agreement with galvanostatic discharging, voltage relaxation on current interruption and impedance measurements. The analysis of the model shows that the overpotential across the LiPON electrolyte is most dominant and is therefore an important rate-limiting factor. It was also found that both ionic diffusion coefficients and electronic conductivity in the $LiCoO_2$ electrode seriously influence the battery performance.

References

1. Bagotsky, V.S., Skundin, A.M., Volfkovich, Y.M. 2015. *Electrochemical Power Sources: Batteries, Fuel Cells, and Supercapacitors*. Wiley, 400 pp.

2. Ragone, D., 1968. Review of battery systems for electrically powered vehicles. *SAE Technical Paper* Article No. 680453.

3. Sun, C., Liu, J., Gong, Y., Wilkinson, D.P., Zhang, J. 2017. Recent advances in all-solid-state rechargeable lithium batteries. *Nano Energy* 33: 363–86.

4. Patil, A., Patil, V., Shin, D.W., Choi, J.-W., Paik, D.-S., Yoon, S.-J. 2008. Issue and challenges facing rechargeable thin film lithium batteries. *Mat. Res. Bull.* 43: 1913–42.

5. Souquet, J.L., Duclot, M. 2002. Thin film lithium batteries. *Solid State Ionics* 148: 375–79.

6. Bates, J.B., Dudney, N.J., Neudecker, B., Ueda, A., Evans, C.D. 2000. Thin-film lithium and lithium-ion batteries. *Solid State Ionics* 135: 33–45.

7. Jones, S.D., Akridge, J.R. 1994. Thin film rechargeable Li batteries. *Solid State Ionics* 69: 357–68.

8. Dudney, N. 2008. Thin film micro-batteries. *Interface* No. 3: 44–8.

9. Liang, C.C., Bro, P. 1969. A high-voltage, solid-state battery system I. *J. Electrochem. Soc.* 116: 1322–3.

10. Liang, C.C., Epstein, J., Boyle, G.H. 1969. A high-voltage, solid-state battery system II. *J. Electrochem. Soc.* 116: 1452–4.

11. Kanehori, K., Matsumoto, K., Miyauchi, K., Kudo, T. 1983. Thin film solid electrolyte and its application to secondary lithium cell. *Solid State Ionics* 9–10: 1445–8.

12. Ohtsuka, H., Okada, S., Yamaki, J. 1990. Solid state battery with Li_2O-V_2O_5-SiO solid electrolyte thin film. *Solid State Ionics* 40: 964–6.

13. Jones, S.D., Akridge, J.R. 1993. A thin-film solid-state microbattery. *J. Power Sources* 43–44: 505–13.

14. Levasseur, A., Kbala, M., Hagenmuller, P., Couturier, G., Danto, Y. 1983. Elaboration and characterization of lithium conducting thin film glasses. *Solid State Ionics* 9–10: 1439 44.

15. Bates, J.B., Dudney, N.J., Gruzalski, G.R., Zuhr, R.A., Choudhury, A., Luck, C.F., Robertson, J.D. 1992. Electrical properties of amorphous lithium electrolyte thin films. *Solid State Ionics* 53–56: 647–54.

16. Bates, J.B., Dudney, N.J., Gruzalski, G.R., Zuhr, R.A., Choudhury, A., Luck, C.F., Robertson, J.D. 1993. Fabrication and characterization of amorphous lithium electrolyte thin films and rechargeable thin-film batteries. *J. Power Sources* 43/44: 103–10.

17. Bates, J.B., Dudney, N.J., Lubben, D.C., Gruzalski, G.R., Kwak, B.S., Yu, X., Zuhr, R.A. 1995. Thin-film rechargeable lithium batteries. *J. Power Sources* 54: 58–62.

18. Wang, B., Bates, J.B., Hart, F.X., Sales, B.C., Zuhr, R.A., Robertson, J.D. 1996. Characterization of thin-film rechargeable lithium batteries with lithium cobalt oxide cathodes. *J. Electrochem. Soc.* 143: 3203–13.

19. Nagasubramanian, G., Doughty, D.H. 2004. Electrical characterization of all-solid-state thin film batteries. *J. Power Sources*, 136: 395–400.

20. Yada, C., Iriyama, Y., Abe, T., Kikuchi, K., Ogumi, Z. 2009. A novel all-solid-state thin-film-type lithium-ion battery with *in situ* prepared positive and negative electrode materials. *Electrochem. Comm.* 11: 413–6.

21. Notten, P.H.L., Roozeboom, F., Niessen, R.A.H., Baggetto, L. 2007. 3-D integrated all-solid-state rechargeable batteries..*Adv. Mater.* 19: 4564–7.

22. Ferrari, S., Loveridge, M., Beattie, S.D., Jahn, M., Dashwood, R.J., Bhagat, R. 2015. Latest advances in the manufacturing of 3D rechargeable lithium micro batteries. *J. Power Sources* 286: 25–46.

23. Long, J.W., Dunn, B., Rolison, D.R., White, H.S. 2004. Three-dimensional battery architectures. *Chem. Rev.* 104: 4463–92.

24. Oudenhoven, J.F.M., Baggetto, L., Notten, P.H.L. 2011. All-solid-state lithium-ion microbatteries: A review of various three-dimensional concepts. *Adv. Energy Mater.* 1: 10–33.

25. Edstrom, K., Brandell, D., Gustafsson, T., Nyholm, L. 2011. Electrodeposition as a tool for 3D microbattery fabrication. *Interface* 20: 41−6.

26. Roberts, M., Johns, P., Owen, J., Brandell, D., Edstrom, K., El Enany, G., Guery, C., Golodnitsky, D., Lacey, M., Lecoeur, C., Mazor, H., Peled, E., Perre, E., Shaijumon, M.M., Simon, P., Taberna, P.-L. 2011. 3D lithium ion batteries—from fundamentals to fabrication. *J. Mater. Chem.* 21: 9876−90.

27. Arthur, T.S., Bates, D.J., Cirigliano, N., Johnson, D.C., Malati, P., Mosby, J.M., Perre, E., Rawls, M.T., Prieto, A.L., Dunn, B. 2011. Three-dimensional electrodes and battery architectures. *Mat. Res. Bull.* 36: 523−31.

28. Rolison, D.R., Long, J.W., Lytle, J.C., Fischer, A.E., Rhodes, C.P., McEvoy, T.M., Bourga, M.E., Lubers, A.M. 2009. Multifunctional 3D nanoarchitectures for energy storage and conversion. *Chem. Soc. Rev.* 38: 226−52.

29. Cohen, E., Menkin, S., Lifshits, M., Kamir, Y., Gladkich, A., Kosa, G., Golodnitsky, D. 2018. Novel rechargeable 3D-microbatteries on 3D-printed-polymer substrates: Feasibility study. *Electrochim. Acta* 265: 690–701.

30. Golodnitsky, D., Nathan, M., Yufit, V., Strauss, E., Freedman, K., Burstein, L., Gladkich, A., Peled, E. 2006. Progress in three-dimensional (3D) Li-ion microbatteries. *Solid State Ionics* 177: 2811–19.

31. Kotobuki, M., Suzuki, Y., Munakata, H., Kanamura, K., Sato, Y., Yamamoto, K., Yoshida, T. 2010. Fabrication of three-dimensional battery using ceramic electrolyte with honeycomb structure by sol–gel process. *J. Electrochem. Soc.* 157: A493–A498.

32. Zhang, F., Wei, M., Viswanathan, V.V., Swart, B., Shao, Y., Wu, G., Zhou, C. 2017. 3D printing technologies for electrochemical energy storage. *Nano Energy* 40: 418–31.

33. Sun, K., Wei, T.-S., Ahn, B.Y., Seo, J.Y., Dillon, S.J., Lewis, J.A. 2013. 3D printing of interdigitated Li-ion microbattery architectures. *Adv. Mater.* 25: 4539–43.

34. Wei, M., Zhang, F., Wang, W., Alexandridis, P., Zhou, C., Wu, G. 2017. 3D direct writing fabrication of electrodes for electrochemical storage devices. *J. Power Sources* 354: 134−47.

35. Yang, Y., Jeong, S., Hu, L., Wu, H., Lee, S.W., Cui, Y. 2011. Transparent lithium-ion batteries. *Proc. Natl. Acad. Sci. U.S.A.* 108: 13013−8.

36. Oukassi, S., Baggetto, L., Dubarry, C., Le Van-Jodin, L., Poncet, S., Salot, R. 2019. Transparent thin film solid-state lithium ion batteries. *ACS Appl. Mater. Interfaces* 11: 683−90.

37. Raijmakers, L.H.J., Danilov, D.L., Eichel, R.-A., Notten, P.H.L. 2020. An advanced all-solid-state Li-ion battery model. *Electrochim. Acta* 330: Article No. 135147.

CHAPTER

2

Materials for All-Solid-State Thin-Film Batteries

2.1 Materials for negative electrodes

2.1.1 Lithium metal

From the point of achieving maximum specific capacity, the optimal material of the negative electrode is plain metallic lithium. In addition, the working (discharge) potential of metallic lithium, in principle, is more negative than the operating potential of all other electrodes, therefore, the use of metallic lithium, with all the other conditions being the same, provides the highest discharge voltage. However, as mentioned earlier, the use of lithium metal as a rechargeable negative electrode in batteries with a liquid aprotic electrolyte encounters the fundamental problems of dendritic formation and encapsulation. Both problems lead to a sharp reduction in cycle-life.

When metallic lithium comes into contact with a solid (crystalline) electrolyte, the problems of dendritic formation and encapsulation do not seem to exist, or in any case, do not play a decisive role. In general, this statement is evidently confirmed by the successful commercialization (or approaching commercialization) of all-solid-state batteries with a metallic lithium negative electrode, implemented in the Oak Ridge National Laboratory, as well as in such companies as STMicroelectronics, CymbetTM Corporation, Front Edge Technology, Inc., Excellatron, and so on. For instance, STMicroelectronics declared the cycle-life of their thin film batteries as high as 4000 cycles. This figure was obtained during not very severe cycling conditions, 1 C discharge rate: cycling between states of discharge 75 and 0% at room temperature. The mentioned battery with nominal capacity 1 mAh uses LiPON as an electrolyte, and $LiCoO_2$ as an active material of positive electrode. CymbetTM Corporation announced 5000 cycles for 10% depth of discharge, and 1000 cycles for 50% depth of discharge.

It is interesting to note that as far back as 1994, it was proposed to apply a thin film of LiPON-like material to a lithium metal electrode to prevent dendrite formation at the boundary of lithium metal with a traditional liquid or polymer electrolyte [1].

As a rule, metallic lithium electrodes are vapor-deposited by thermal evaporation or RF and DC magnetron sputtering on mechanically and chemically stable solid electrolyte [2, 3]. The typical thickness of lithium thin-film electrode amounts to 2–5 µm, which corresponds to capacity of 0.4–1.0 mAh/cm².

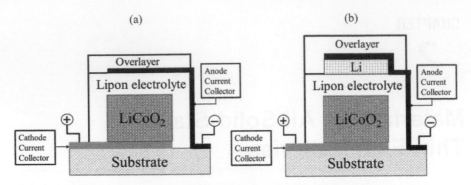

Figure 2.1. Schematic cross-sectional views of a Li-free thin-film battery with in situ plated Li anode (a) completely discharged and (b) charged

A serious disadvantage of batteries with a metallic lithium negative electrode is the restricted operating temperature, determined by the melting point of lithium, 180.54 °C. Insofar as thin-film all-solid-state batteries are intended first of all for microelectronic devices, such as smart cards, RFIDs and so on, these batteries must be involved in modern semiconductor technology, i.e., must be suitable for soldering at higher temperatures. A rather ingenious version that allows solving this problem is the so-called 'Li-free construction' [4]. Such a lithium-free battery is assembled without lithium metal (and therefore, it survives high-temperature solder reflow procedures) but with some excess of active material of positive electrode. In the course of the first charge the proper amount of lithium is plated at the current collector (Fig. 2.1). For the proper functioning of the lithium-free battery, it is very important that the material of the current collector in the negative electrode does not form intermetallic compounds with lithium. The most suitable material in this respect is copper, which is used as a current collector in conventional lithium-ion batteries, although some alternative materials, such as Ti, Co and TiN are discussed as well.

An unusual version of lithium-free battery is described in [5]. This battery is assembled from two current collectors and the electrolyte (LiPON) sandwiched in between. It is LiPON that is the only source of lithium for both electrodes. The current collector of the positive electrode is made from silver, whereas the current collector of the negative electrode is made from stainless steel. Obviously, the battery is assembled in the discharged state. In the course of the first charge some amount of silver is anodically dissolved, that enters the LiPON structure and replaces lithium ions. Simultaneously lithium ions are cathodically deposited onto a stainless steel current collector, forming lithium negative electrode. Although the lithium-free battery seems to be attractive, this approach has not received much development.

Another approach to increasing the operating temperature of thin-film all-solid-state batteries is to replace pure lithium with a lithium alloy, for example, alloy with magnesium. The melting point of such an alloy, depending on the content of magnesium, varies from 200 °C with a magnesium content of 4 at. % up to 400 °C with a magnesium content of 40 at. %.

A more fundamental solution to the problem is, however, replacing the lithium electrode with an electrode typical for lithium-ion batteries, i.e., with the electrode, where lithium ions are reversibly inserted in some matrix. As such matrices (as in traditional lithium-ion batteries), elements of the fourth group of the periodic table (carbon, silicon, germanium, tin), oxides and some other compounds can be used. Such electrodes are described here and later.

2.1.2 Silicon and silicon-based composites

Silicon has a record capacity for the reversible insertion of lithium. When lithium is inserted into silicon, intermetallic alloys are formed, and the lithium-richest intermetallic compound is $Li_{4.4}Si$ ($Li_{22}Si_5$), which corresponds to a specific capacity of 4200 mAh/g. Such an alloy is formed only at elevated temperatures. At room temperature, the lithium-richest intermetallic compound is $Li_{3.75}Si$ ($Li_{15}Si_4$), which corresponds to a specific capacity of 3590 mAh/g. It should be emphasized that the specified values of a particular specific capacity refers to the process of lithium insertion into silicon, i.e., to the charge of the negative electrode. When discharging, i.e., when extracting lithium from the $Li_{3.75}Si$ intermetallic compound, the specific capacity is 1852 mAh/g (let us remember that the specific capacity of pure lithium is 3828 mAh/g). As indicated earlier, in modern lithium-ion batteries, negative electrodes are made of graphite. The maximum amount of lithium that can be inserted (intercalated) into graphite corresponds to the composition of LiC_6, i.e., the specific intercalation capacity of graphite amounts to 372 mAh/g. Therefore, silicon has a specific capacity of an order of magnitude higher. A number of reviews devoted to application of silicon in lithium-ion batteries can be found [6–11].

According to the equilibrium phase diagram, several intermetallic phases exist in the system Li–Si from LiSi to $Li_{22}Si_5$. However, during lithiation of silicon at room temperature, it is not possible to observe the corresponding steps on the charge curve. At the first cathodic polarization of the crystal silicon its progressive amorphization takes place (i.e., amorphous intermetallic compounds are formed). Following anodic polarization (delithiation) the silicon is left amorphous. Typical charge (cathodic) and discharge (anodic) curves taken in thin-film silicon electrodes are shown in Fig. 2.2.

Lithium insertion into silicon proceeds at potentials close to that of lithium, whereas lithium extraction from lithiated silicon takes place in the potential range 0.3 to 0.5 V *vs.* Li^+/Li.

When lithium, density of which is only 0.5 g/cm^3, is inserted into a matrix of foreign material (usually having a higher density), an increase in specific volume occurs, which is associated with the creation of internal stresses. With complete lithiation of carbon, when only one lithium atom falls on six carbon atoms, the increase in specific volume is small, does not exceed 12%. When lithium is inserted into silicon, where there are almost four lithium atoms per silicon atom, the specific volume increases to threefold. The resulting internal stresses cause complete destruction of monolithic silicon up to pulverization. Simultaneously, the silicon loses contact with a current collector, resulting in a huge Ohmic drop. Figure 2.3. illustrates the cracking of the silicon plate and its surface destruction (some kind of 'etching') at cycling in 1 M $LiClO_4$ in a mixture of propylene carbonate and dimethoxyethane.

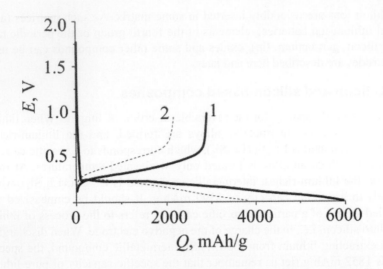

Figure 2.2. Typical charge and discharge curves at thin-film silicon electrodes for 1st and 2nd cycles

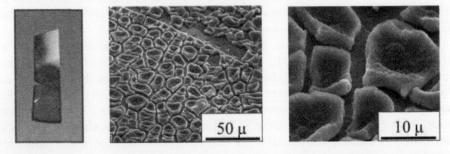

Figure 2.3. Photo of silicon plate after 2 cycles with current 5 mA/g (left), and SEM images of the surface with two different magnification (center and right)

At the end of the 20th century, it was found that silicon-based nanostructured materials are resistant to mechanical destruction with repeated electrochemical insertion of lithium from traditional aprotic electrolytes. Initially, the most impressive successes were achieved on samples with thin films of amorphous silicon [12, 13]. Thus, [14] provides data on the cycling of electrodes with an amorphous silicon film 50 nm thick for 100 cycles with a specific capacity of 3500 mAh/g. In [13] more than 1000 cycles with a capacity of 1500 mAh/g on a film of amorphous silicon with a thickness of 77 nm were obtained. [15] reported that a 2000 mAh/g capacity was obtained over 180 cycles on a 300 nm thick film. At that time (the beginning of the 21st century), it was found that the cyclability of amorphous silicon films decreases markedly with increasing film thickness. Thus 1000 cycles with a capacity of more than 3000 mAh/g were obtained on films with a thickness of 50 nm, while 200 cycles with a capacity of 2500 mAh/g were obtained on a film with a thickness of 150 nm

[16]. On the 340 nm thick films, 450 cycles with a capacity of 2000 mAh/g were obtained, whereas only 50 cycles with the same capacity were reported on a 3.6 μm thick film [17]. In this regard, thin silicon films on metal substrates observe the general Griffith-Irwin relation, connecting the film thickness d and critical fracture stress σ:

$$\sigma = K/\sqrt{\pi d} \tag{2.1}$$

here K is the fracture toughness of the film material.

The physical meaning of the Equation (2.1) conforms to the fact that an individual crack can only be initiated, if the driving force of the crack's appearance is larger than the cracking-resistant force. Therefore, cracking is possible if the stored strain energy exceeds the energy required for generating a new surface of the crack. Thus, there is some critical film thickness below which cracks cannot form in principle. Different works have point out that this thickness for silicon films falls in the range from 50 to 300 nm.

Later the intensity of research on thin-film silicon electrodes was significantly increased, see e. g., reviews [18, 19].

Conventionally, the definition of a 'thin film' is attributed to objects with a thickness from tens of nanometers to several micrometers. A high surface area to volume ratio inherent for thin-film electrodes, provide quite fast lithiation and delithiation due to a very small diffusion length and a short pathway for electron transfer. In other words, such a design provides a rather high C-rate performance. Low thickness of thin-film electrodes promotes and also accommodates mechanical stresses induced by volume changes on lithium insertion and extraction. Once more attractive feature of silicon-based thin-film electrodes consists on the fact that they do not need in a conductive additives and binders.

Among the various factors that cause the degradation of thin silicon films during cycling, the insufficient adhesion of these films to the substrate plays a significant role. It was found that increasing the surface roughness of the substrate significantly improves the cyclability of thin-film electrodes. Indeed, the surface roughness increases the contact area and enhances the adhesion. Furthermore, modifications of the substrate surface can create mechanical anchor points for the deposited silicon film, providing a continuous electrical contact of the film with the substrate even when film cracking has occurred. Various methods of roughening copper (and nickel) substrates are described, including mechanical abrasion with sandpaper (from 100 to 260 nm roughness), and etching with an aqueous solution of $FeCl_3$ (0.2–2.0 M). In some cases, such roughness of the surface of the substrate is artificially created, for example, by the deposition of pyramid-like tiny hills of copper onto a copper substrate [20]. A rather sophisticated, but very effective method of surface modification is a treatment with lanthanum plasma immersion ion implantation [21].

In general, the substrate (current collector) has a notable effect on the characteristics of electrodes with thin layers of active material. Very often, copper foil is used as a current collector for silicon films. Since copper has a fairly high density (about 8.9 g/cm³), the total weight characteristics of the thin-film electrode happen to be quite low. In this regard, the proposal to replace the copper current collector with a network of carbon nanotubes [22] is of great interest. Here, silicon is deposited into the flexible conducting network of carbon nanotubes with the formation of thin free-standing electrode with good cyclability.

In addition to thin films, nanosheets can also be attributed to 2D materials. Interest in these materials with respect to lithium-ion batteries began in 2011. Certain advantages of silicon nanosheets consist in easy accessibility of Li-ions, as well as in some tolerance toward volume change at lithium insertion/extraction. Initially, the electrodes with ultrathin silicon nanosheets were prepared by using nanosheets of graphene oxide as a template [23]. Later other materials, including nickel nanosheets were used as a substrate for silicon nanosheets.

When lithium is inserted into silicon electrodes (as well as when it is inserted into other matrices), the electrode acquires a close potential to that of lithium, i.e., negative enough for the inevitable reduction of electrolyte components, both solvent and lithium salt. This process proceeds at the first cathodic polarization (the first charging of the negative electrode) and results in unproductive charge consumption, i.e., in irreversible capacity loss. The electrolyte reduction results in formation of insoluble products, which are deposited on the surface of the negative electrode in the form of a passive film, known as 'solid electrolyte interphase', SEI. The SEI has a notable ionic conductivity, and thus prevents the direct contact of silicon with a liquid electrolyte, and therefore further reduction of the electrolyte.

The conditions for the SEI formation (or the conditions for the reduction of electrolyte components) on silicon differ from the corresponding conditions on carbon. The electrolyte is reduced on silicon at much more negative potentials than on carbon, usually at potentials more negative than 0.6 V. It is known that, initially, the silicon surface is always covered with a kind of native film, which can contain both oxide groups (Si – O – Si) and silanol (Si – OH). Such groups themselves can be reduced with the formation of the SEI, which makes a definite contribution to irreversible capacity. In practice, already in the first cycles, the native film is completely replaced by the SEI, which is characteristic of these conditions. Naturally, the conditions for the formation of the SEI on silicon depend on the electrolyte composition. In the simplest case, the SEI has a mosaic structure and consists of organic and inorganic salts (Li_2CO_3, LiF, LiOCOOR) and materials like polyethylene oxide. The SEI on silicon has a dynamic nature (as on carbon) and its structure changes during cycling [24]. A significant improvement in the SEI (and accordingly, irreversible capacity diminution) can be achieved by introducing various additives into the electrolyte. These are large volume changes and the formation of unstable SEI film are the main issues for the failure of silicon-based electrodes.

As it was noted early a very large volume expansion corresponds to the emergence of associated stress. This stress can be rather high, up to 3 GPa. This results in cracking and fractures of the rigid silicon film, therefore in the appearance of a fresh silicon surface. New SEI is required for this new surface, and the process of SEI formation consumes lithium from the positive electrode, and therefore decreases the cell's capacity.

To prevent or at least reduce the formation of SEI in cracks and pores of silicon films, it was proposed to apply specific coatings on such films, in particular, coating from fullerenes. [25] describes the polymeric coating from fullerene C_{60}, deposited by a plasma-assisted thermal evaporation technique to surface of 300 nm thick silicon films. The discharge capacity of bare silicon film at cycling with current density of 0.1 mA/cm^2 decreased from 3000 to 250 mAh/g for 50 cycles, whereas the capacity of Si film with 100 nm polymeric fullerene coating retained at a level no less than 3000 mAh/g in these conditions.

Thin films are an example of 2D nanomaterials. It is clear that such materials are of great interest for all-solid-state thin-film batteries. There are a variety of techniques for deposition of thin films of silicon-based materials, and the majority of these methods fall into two main categories, namely Physical Vapour Deposition (PVD) and Chemical Vapour Deposition (CVD). PVD consists in evaporation of solid initial material and subsequent condensation of the vapor at substrate surface. Plain or doped silicon can be a source material for PVD of silicon films. For creating composite or multi-layered films, a collection of different materials play a role of source materials. Various methods are used for heating and evaporation of the source material, including direct thermal evaporation, pulsed laser evaporation, electron-beam evaporation. Probably, the most popular method of thin films deposition is magnetron sputtering in both Direct Current (DC) and Radio Frequency (RF) versions. A significant advantage of magnetron sputtering is good adhesion of the deposited material to the substrate as well as a possibility to deposit uniform films onto large-sized substrates. As a rule, PVD with thermal evaporation is based on electric heating of the target. This technique provides a rather high deposition rate, up to several micrometers per minute. Pulsed laser deposition, which is based on target material evaporation due to the action of a focused pulsed laser beam, gives a possibility to prepare the high-quality films with the same stoichiometry as the target. Usually this method is used for deposition of materials with high melting temperatures. PVD can be realized at room or at elevated temperatures. Thin films of silicon-based materials produced by PVD, are usually amorphous. PVD techniques give a possibility to precisely control thickness and morphology of thin films.

The CVD method is based on chemical reactions, namely decomposition of gaseous silicon-containing precursors under vacuum. Silane (SiH_4), disilane (Si_2H_6) and trichlorosilane ($SiHCl_3$) are used as silicon precursors most frequently. CVD produces high-quality thin silicon films. The temperature of the CVD process can vary from 150 to *ca.* 1000 °C, the films produced at low temperatures are amorphous, whereas those produced at high temperatures are nanocrystalline.

It is worth mentioning that one more method of thin film deposition exists, namely electrodeposition of silicon from non-aqueous electrolytes. Until very recently this method remained somewhat unusual. Works devoted to such a method are few and far between.

It is well known that in many cases, composite electrodes have certain advantages over single-component electrodes. This situation is true for electrodes based on silicon, including thin-film ones. Among them, the most interesting are multilayer electrodes, in which the active part consists of many alternating thin layers of silicon and other material. This other (secondary) material can be carbon, silver, iron, germanium, and even porous (so-called 'low-density') silicon [26–31]. In the last case low-density layers play a role of some kind of restriction that relaxs the stress and prevents crack propagation provoked by lithium insertion/extraction. Probably, the best version of multi-layered electrodes is based on the system Si-O-Al

[32]. Such electrodes contain the active parts with a thickness up to 3 μm prepared by alternate magnetron sputtering of pure silicon and a mixture of silicon and aluminum in an oxygen-containing environment (Fig. 2.4). The electrodes demonstrate a sufficiently good cyclability over more than 200 charge-discharge cycles with the capacity fading less than 0.09% per cycle.

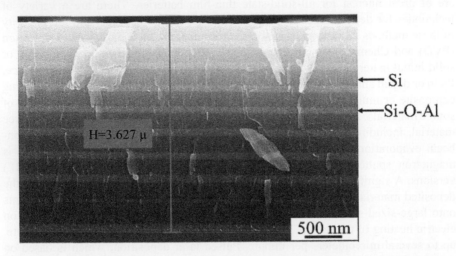

Figure 2.4. SEM image of the cleavage of the fresh-prepared Si–O–Al composite film on the titanium foil. Light bands refer to layers enriched in aluminum, dark bands refer to layers depleted in aluminum

Figure 2.5 shows charge and discharge curves taken at such multilayer electrodes. These curves differ but insignificantly from those taken at plain silicon electrodes.

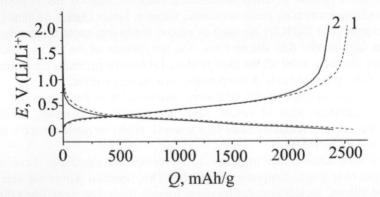

Figure 2.5. Charge-discharge curves at multilayer Si–O–Al composite electrode at rate C/12 (curve 1) and C/3 (curve 2)

Also, of interest are thin film electrodes made from an amorphous silicon matrix in which phosphorous doped nanocrystalline silicon is embedded. These films are synthesized by the Plasma Enhanced Chemical Vapor Deposition (PECVD) method exhibiting excellent electronic conductivity (up to 35 S/cm) and show high discharge capacity of 2250 mAh/g for 50 cycles at C/5 rate.

Besides 2D silicon nanomaterials, their 0D, 1D and 3D counterparts have also attracted the attention of modern researchers [7, 33, 34]. Nanowires (nano whiskers) [35] and nanotubes [36] are examples of 1D materials with quite good cyclability.

Some mathematical model relating surface energy with diffusion-induced stresses in nanowire electrodes was developed in [37]. Insofar as the ratio of surface area to volume increases with decreasing nanowire diameter, surface energy influences such stresses in nanostructured electrodes. The model establishes a relationship between stress and the nanowire diameter (the less the diameter, the lesser the mechanical degradation).

Silicon nanowires can be synthesized by growth or etching methods, the former being the most popular. Growth methods are usually based on the so-called Vapor–Liquid–Solid (VLS) mechanism in the version of Chemical Vapor Deposition (CVD). In this case gas-precursors are silane (SiH_4) or silicon tetrachloride ($SiCl_4$), although intermediate species, such as disilane (Si_2H_6) and dichlorosilane (SiH_2Cl_2) can be used as well. Depending on the precursor's nature the temperature of CVD process can vary in quite a wide range, from 300 to 1000 °C. Chlorinated silanes are known to be more chemically stable, and they demand higher temperature for CVD realization. Conditionally, CVD of silicon nanowires can be considered as high-temperature processes (800–1000°C by using $SiCl_4$), and low-temperature process (300–600°C by using SiH_4).

High-temperature CVD of silicon nanowires is carried out in flow-through reactors in the form of tubular oven with external heating. Gas-carrier (hydrogen or mixture of hydrogen with inert gas) is passed through a bubbler with liquid $SiCl_4$ just prior to entering the tubular oven. A substrate with an array of catalyst droplets is placed preliminary in the hot zone of the reactor. The catalyst is a key-point of the CVD process. Very often such a catalyst is a metal, forming simple eutectic with silicon, e. g., gold, silver, copper. At the first contact of the catalyst (e.g., Au) drop with Si-bearing gas, the thermal decomposition of $SiCl_4$ takes place with the formation of liquid eutectic Au-Si. As $SiCl_4$ decomposes further, eutectic oversaturation and solid silicon precipitation occur. As a result, quite long, whisker-like wires with a spherical cap are formed. The morphology of nanowires depends on the synthesis conditions. In tubular reactors with external heating (when the temperature of the reactor wall is higher than the temperature of the sample inside the pipe), a 'forest' of parallel nanowires grows perpendicular to the surface of the substrate. Such nanowires have a very low defect density. In reactors with internal heating, when the temperature of the sample exceeds the temperature of the walls of the reactor, curved and kinked nanowires form, creating a highly porous mat layer on the surface of the substrate. It is worth noting that the crystal structure of nanowires depends on their diameters. For nanowires with diameter less than 20 nm, the <110> growth direction is mostly encountered, whereas for nanowires with diameter above 50 nm, the <111> direction is favored. For nanowires with diameters from 20 to 50 nm the <112> direction is often observed.

Like thin films, silicon nanowires can be electrolytically deposited. For this end, a solution of $SiCl_4$ in ionic liquid can be used as an electrolyte. Electrodeposition is carried out using a mask from the ion-track etched polycarbonate membrane with pores diameter from 15 to 400 nm. This technique allows producing nanowires with strongly identical diameters.

Often silicon nanowires have a so-called 'core-shell' structure with an inner core from a crystal silicon and an outer shell from an amorphous one. In this case crystal core provides sufficient electron conductivity, whereas the amorphous shell is used for storing lithium. Other version of nanowires with core-shell structure have also been described, the most important being nanowires with carbon cores (carbon nanofibers or carbon nanotubes) and shells from amorphous silicon. Such materials allow deeper lithiation, because there is no danger of core amorphization in this case. Unfortunately, carbon nanofibers are prone to embrittlement on lithiation.

Another approach to composite nanowires is silicon nanowires with conductive coating. Such a coating should be permeable for lithium ions and usually this coating is made from carbon [38]. Silicon nanowires with metal coatings (copper, aluminum) are also described.

Another representative of 1D materials is silicon nanotubes. In principle, hollow nanotubes must be able to limit volume changes during lithiation/delithiation. Although silicon nanotubes seem to be quite attractive, there are only few papers describing their use in lithium-ion batteries. Silicon nanotubes were synthesized by template techniques.

A lot of research was carried out with silicon powders or particle-based structures which are 0D materials. It is clear that such structures must be nano-sized, because for the above mentioned large volume expansion during lithiation which result in cracking and crushing of Si particles. It was documented (see, e. g., [39]) that some critical (threshold) particle's diameter exists, below which a silicon particle does not crack. This threshold was estimated at about 150 nm. In reality, such particles should be even smaller. Indeed, although sub-threshold-sized particles do not crack, they lose or worsen an electrical contact with the current collector due to periodic volume expansion and shrinkage in the course of lithiation/delithiation.

Simple silicon nanopowders with particle size 50 or 100 nm are commercially available now. They can be used in negative electrodes, but their electrode behavior is not the best. Much better are silicon hollow nanoparticles. Such a material could be prepared by a template technique from silica nanopowder (particles with 350 nm in diameter). A thin layer of silicon is deposited onto silica nanosphere by the CVD method from SiH_4. The thickness of the layer can be as small as 10 nm. Then the silica core is etched out with dilute hydrofluoric acid. The electrodes based on such hollow nanospheres demonstrated discharge capacity up to 2725 mAh/g and were tested for 700 cycles with modest capacity loss. It is the presence of void spaces in hollow nanoparticles that provides good cyclability of the electrodes. Besides, such hollow nanoparticles have stable SEI layers and reveal reliable contact with current collector and particles of conductive additive.

The next step in improvement of 0D silicon materials is a core-shell structure. The simplest and most common example of such a material is carbon coated nanoparticles of solid silicon. For example, Ref. [40] describes silicon nanoparticles with a size of less than 100 nm coated with amorphous carbon with a thickness of up to 10 nm. To apply such a coating, a suspension of silicon nanopowder in an aqueous solution of polyvinyl alcohol was dried and then subjected to heat treatment at a temperature of 750 °C until the polyvinyl alcohol was completely carbonized. The electrodes from the obtained material had a reversible capacity of about 1800 mAh/g for 50 cycles without notable degradation. The presence of an electron-conducting shell not only ensures reliable electrical contact between the particles, but also prevents their agglomeration. Other different methods for applying a carbon shell to silicon nanoparticles are described as well.

More sophisticated versions of silicon-based core-shell structures are also defined. For example, a double-walled core–shell structure was prepared and investigated in [41]. In this case, Si nanoparticles (100–200 nm in diameter) were first calcined in air at the temperature of 700 °C with formation of inner shell from SiO_2 (Si@SiO_2). Second, a certain amount of epoxy resin was heat-treated at 100 °C and mixed with Si@SiO_2 with formation of a gel. Then this gel was calcined in an argon atmosphere at 700 °C to get double-walled core-shell structured Si@SiO_2@C nanocomposite. The electrodes from this nanocomposite showed a reversible specific capacity of about 786 mAh/g after 100 cycles at a current density of 100 mA/g with a degradation rate of 0.13% per cycle. Such good electrochemical performance was explained by the double walls of carbon and SiO_2 enhanced the electronic conductivity and the compatibility of electrode materials and the electrolyte as a result of accommodating the significant volumetric change during cycles. The authors believed that the interlayer SiO_2 may release the mechanical strain and enhance the interfacial adhesion between the carbon shell and silicon core.

Besides carbon, other materials including copper and titania can be used as shells in silicon-based core-shell structures.

It seems that core-shell structures with hollow cores could have advantages over structures with solid cores. However, synthesis of hollow core-shell structures is rather difficult, and only a limited number of corresponding references can be found in the literature. One of such examples is described in [42]. In this work hollow silicon nanoparticles with silver shells were prepared by multi-step synthesis. Initially, nanoparticles of polystyrene (120 nm in diameter) were coated with sol-gel SiO_2. Then this material was heat-treated at 600 °C to burn out polystyrene with formation of hollow silica nanoparticles. By a magnesiothermic reduction these nanoparticles were transformed into monodisperse hollow silicon nanospheres. Finally, these nanospheres were coated with a carbon shell (HPSi@C) or silver shell (HPSi@Ag). The electrodes based on HPSi@Ag composites demonstrated a discharge capacity of about 3000 mAh/g for 100 cycles at the current 0.5 A/g with a very low capacity loss. At the current 4 A/g their capacity was about 2000 mAh/g.

Of late, the so-called yolk-shell structures have gained great popularity. The yolk-shell structure has a certain void in between the solid silicon core and outer conductive shell. It turned out that the synthesis of the yolk-shell structure is simpler than that of hollow core-shell one. The yolk-shell structure compensates the volume changes of the silicon core by the so-called engineered void (Fig. 2.6). At the same time, shell from conductive material provides not only reliable electron transport, but the stability of the SEI layer as well. The typical synthesis of the yolk-shell structure is the following. At first, silicon nanoparticles are coated with SiO_2 by usual sol-gel technique. Then such Si@SiO_2 particles are coated with carbon by sucrose

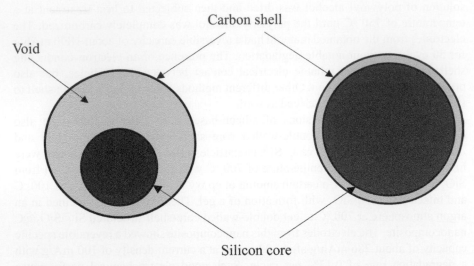

Figure 2.6. The functioning principle of the yolk-shell Si/C hierarchical structure design

carbonization at 800 °C. Finally, the silica is etched out with hydrofluoric acid resulting in the formation of the Si@void@C yolk–shell structure. Unfortunately, the performances of electrodes based on the Si@void@C yolk–shell structure are inferior to performances of above described HPSi@Ag-based ones.

Like the above-described double-walled core–shell structures, certain dual yolk–shell structures (Si@void@SiO$_2$@void@C) are mentioned in the literature.

2.1.3 Lithium titanate

Lithium titanate Li$_4$Ti$_5$O$_{12}$, as a material of an electrode functioning on the principle of reversible lithium insertion, was mentioned in the literature as far back as the 80's of the last century, i.e., long before the idea of lithium-ion batteries [43, 44]. Sometime later, the use of lithium titanate (LTO) as the active material of negative electrodes in lithium-ion batteries was patented [45]. A salient feature of this material is the practical constancy of its specific volume when fully lithiated (and back delithiated), and, as a result, the absence of internal stresses during the electrode cycling [46]. The lithiation process is described by the Equation

$$\text{Li}_4\text{Ti}_5\text{O}_{12} + 3e^- + 3\text{Li}^+ = \text{Li}_7\text{Ti}_5\text{O}_{12}, \tag{2.2}$$

therefore, the theoretical specific capacity of this process is 175 mAh/g. Lithium titanate Li$_4$Ti$_5$O$_{12}$ is crystallized in the spinel structure, and as a result of reaction (2.2), the initial structure of a substance of the spinel type transforms into a metastable phase with an ordered structure of the NaCl type. At the end of the last century, this fact was proved by X-ray diffraction analysis on samples with various degrees of lithiation. Thus, like lithium iron phosphate, and unlike most other materials for lithium-ion battery electrodes, the potential of which depends on the lithiation degree (the coefficient x in the formula of the active substance, for example, Li$_x$CoO$_2$), in this case, two-phase equilibrium (Li$_4$Ti$_5$O$_{12}$)^{spinel}/(Li$_7$Ti$_5$O$_{12}$)$^{\text{quasi-NaCl}}$ exists, and in the course of lithiation and delithiation, the relative amounts of phases of the starting material and reaction product changes, whereas the equilibrium potential remains constant regardless of the degree of lithiation. According to many authors, the value of this equilibrium potential is 1.55 V relative to the lithium electrode. Already in early works on titanate electrodes, practically horizontal charge-discharge curves obtained at low currents were shown. In these works, it was concluded that lithium-titanium spinel is a promising candidate for the negative electrode of lithium-ion batteries [47–49]. Typical charge and discharge curves of the LTO-based electrode are shown in Fig. 2.7.

The picture presented in Fig. 2.7 refers to the low C-rate and the galvanostatic curves that nearly approach the equilibrium ones. They have practically horizontal segments corresponding to a two-phase transition (Equation 2.2) as well as chamfered ones corresponding to rather narrow areas of solid solutions. The increase in C-rate as well as decrease of temperature result in violation of the equilibrium, widening of solid-solution-areas, shortening two-phase segment and diminishing the real capacity. The single-phase ranges at the beginning and the end of charge or discharge depend closely on the LTO particle size, this effect being manifested on going from dozens of nanometers to units of nanometers.

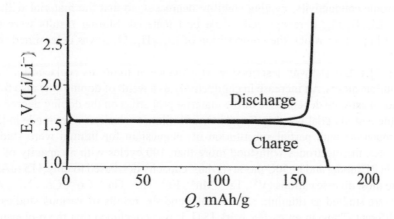

Figure 2.7. Typical charge and discharge curves at LTO-based electrode at C-rate C/10

Since the working potential of reaction (2.2) it was noticeably more positive than the potential of lithium or graphite electrodes, when using LTO as a negative electrode, the battery voltage will be significantly lower than when using a traditional carbon electrode, however this drawback is atoned by the unique cyclability of the material. Since during the transition $Li_4Ti_5O_{12} \rightarrow Li_7Ti_5O_{12}$ volume changes are insignificant, no more than 0.1%, this is considered as a factor contributing to cyclability, since mechanical degradation of the macrostructure of the electrode is excluded. Of great importance for the market prospects of the material is also the low cost of raw materials. It is also very important that the charged form of the LTO ($Li_7Ti_5O_{12}$) is completely safe, unlike LiC_6 and, especially, metallic lithium, since electrolyte reduction does not occur at working potentials, and the probability of dendrite formation even at overcharge is extremely low. Finally, it is worth noting that LTO is environmentally benign.

The most significant disadvantage of lithium titanate as an active material is its low electronic conductivity. The reason for the low electronic conductivity is the absence of charge carriers in the structure of this substance. Since the main metal ion in the composition of the substance is in the Ti^{4+} state and has the $3d^0$ electronic configuration, the conduction band is completely empty. According to various sources [48, 49], the electronic conductivity of $Li_4Ti_5O_{12}$ under normal conditions lies in the range from 10^{-13} to 10^{-8} S/cm, the former figure being more reliable. The creation of the following materials was used as a way to increase the electrical conductivity of LTO: 1) different doped materials; 2) composites with conductive additives; 3) materials with conductive coatings (of which carbon is the most popular) and, finally, various nanomaterials.

Doping of lithium titanate is possible when both titanium or lithium are replaced by foreign cations. Moreover, it was shown in [50] that the substitution of a certain number of titanium atoms by lithium, i.e., the creation of a material with the general formula $Li_{4+x}Ti_{5-x}O_{12}$ ($0 < x < 0.2$) leads to a notable increase in the electronic conductivity of the material (and a corresponding improvement in the operation of the electrode at high rates). However, with an increase in x, along with an increase in electronic conductivity, cycling stability decreases, so that the material with the formula $Li_{4.1}Ti_{4.9}O_{12}$ is recognized as the best material. Similar results were also obtained in [51], in which the composition of $Li_{4.07}Ti_{4.93}O_{12-\delta}$ was determined to be optimal.

In [52], a material was described in which calcium atoms are replaced for some of the lithium atoms. An increase in conductivity as a result of doping was also found here and an extreme dependence of the material properties on the doping degree was noted: the best material was $Li_{3.9}Ca_{0.1}Ti_5O_{12}$. Similar conclusions were made in [53], where materials with partial substitution of magnesium for lithium were studied. In this case, the electrodes withstood more than 100 cycles with a capacity of 170 mAh/g in 2C mode, and in 50C mode (!) their capacity decreased to only 115 mAh/g.

The most diverse cations (V^{5+}, Ta^{5+}, Mn^{4+}, Fe^{3+}, Al^{3+}, Ga^{3+}, Co^{3+}, Cr^{3+}, Ni^{2+}, Ag^+, Mg^{2+}) were studied as titanium substituents, and the results of various studies are quite different. Thus, in an earlier work [54], it was concluded that the replacement of titanium with vanadium is accompanied by a significant decrease in the discharge capacity, whereas in [55] it was shown that the $Li_4Ti_{4.95}V_{0.05}O_{12}$ material retains the same capacity as the undoped sample, but has much less polarization resistance and provides better cyclability of the electrode. A similar conclusion was also made in [56].

Doping of LTO with zinc [57, 58] contributes to an increase in both electronic and ionic conductivity. As a result, such a doped material has quite acceptable performances even with a characteristic particle size that is not too small. The material of the composition $Li_4Ti_{4.8}Zn_{0.2}O_{12}$ showed a discharge capacity of 186 mAh/g after 200 cycles even in the 5C mode. The material $Li_2ZnTi_3O_8$ was described in [59–62], which can be considered, not as zinc-doped-lithium-titanate, but as double spinel $Li_4Ti_5O_{12} - Zn_2TiO_4$.

The beneficial effect of replacing a part of titanium with tungsten was shown in [63]. It is established here that such substitution leads to a significant increase in electronic conductivity. For example, a material of the composition $Li_4Ti_{4.9}W_{0.1}O_{12}$ had a specific conductivity of 0.15 S/cm, which was approximately 12 orders of magnitude higher than the conductivity of undoped lithium titanate. Electrodes with $Li_4Ti_{4.9}W_{0.1}O_{12}$ showed a specific capacity of 128 mAh/g after 100 cycles in 10C mode.

Doping of LTO with aluminum is described in [64–69]. Replacing only 5% titanium with aluminum led to a marked improvement in electrode cycling at higher current densities. Doping of LTO with iron, and especially, with gallium facilitates increased specific capacity at high rates [70, 71].

In [72], a material in which nickel ions simultaneously replace both titanium and lithium atoms were described. The conductivity of a material with the formula $Li_{4-2x}Ni_{3x}Ti_{5-x}O_{12}$, was found to grow with increasing doping degree x to 0.075, while the diffusion coefficient of lithium reaches a maximum at x = 0.05.

An interesting case of joint cation-anion doping was described in a work [73]. It is shown here that treatment of commercial lithium nanotitanate (with a characteristic particle size of 22 nm) by aluminum fluoride leads to a partial replacement of titanium by aluminum and oxygen by fluorine by a simplified reaction

$$Li_4Ti_5O_{12} + AlF_3 \rightarrow TiO_2 + Li_{4+x-y}Ti_{5-x}Al_xO_{12-y}F_y \qquad (2.3)$$

Such substituted titanate had a discharge capacity of 100 mAh/g in 50C mode. In addition, it was concluded that doping with fluorine helps protect titanate from exposure to hydrogen fluoride (a product of the hydrolysis of lithium hexafluorophosphate with traces of moisture in the electrolyte), and thus, increasing the stability with prolonged cycling. Samples of lithium titanate with the replacement of part of the oxygen by fluorine were also studied in detail in [74]. An extreme dependence of the characteristics of the material on the degree of doping was also found here, and the material with the formula $Li_4Ti_5O_{11.7}F_{0.3}$ was recognized as the best.

When creating composites of LTO with electrically conductive additives, different carbon materials, as well as some metals, oxides, and other compounds, were used. The carbon source in the synthesis of composites can be both elemental carbon in the form of graphite, carbon black and various nanomaterials (nanotubes, fullerenes, etc.,), and organic compounds that undergo pyrolysis during synthesis; sucrose being a favored carbon source. Compared with metal conductive additives, carbon has a lower density and a lower cost. Besides increasing the electrical conductivity, the incorporated carbon nanostructures also help control the nanoparticles growth and suppress nanoparticles aggregation. The simplest way to create LTO-carbon composite is carbon coating of LTO particles [75].

For the first time, lithium titanate composites with multi-walled carbon nanotubes were described in [76]. Carbon nanotubes are known to be characterized by unusually high electronic conductivity in the direction of the axes of the nanotubes, therefore, the combination of carbon nanotubes with powder active material provides sufficient conductivity throughout the entire thickness of the porous electrode, and the content of carbon nanotubes in the composite may not be very large. In [76], the content of nanotubes was 8.2%. When creating such composites, it is very important to ensure good adhesion of titanate nanoparticles to the surface of carbon nanotubes. The $Li_4Ti_5O_{12}$ composites with multi-walled carbon nanotubes described in [77], when discharged in the 10C mode, had a capacity of 143 mAh/g. Similar results were also obtained in [78–80]. In the last two cases, the electrodes stably functioned even in the 30C mode. The review [81] reports on the urchin-like composite of $Li_4Ti_5O_{12}$ with carbon nanofibers synthesized by a sol-gel method which delivered a capacity of 122 mAh/g after charging/discharging at 10 C for 500 cycles. Even better rate performances reveal coaxial nanocables with cores from carbon nanotubes and shells (sheaths) from LTO. The thickness of the shells is as low as 25 nm [82].

Another carbon nanoform that has recently attracted the attention of researchers is free graphene. Composites of LTO with graphene are described in a lot of papers. In [83], composites of LTO nanofibers obtained by electrospinning with graphene were prepared. Nanofibrous lithium titanate per se has advantages over conventional powder. The introduction of graphene into this material results in a significant increase of the electronic conductivity and, accordingly, improvement of the operation at high currents. For instance, electrodes based on plain LTO powder, LTO nanofibers, and LTO nanofibers/graphene composite at cycling in 22C mode demonstrated specific capacity, correspondingly, 30, 45 and 110 mAh/g.

In [84], composites of almost spherical LTO nanoparticles with graphene were described. Electrodes with such material were characterized by even better characteristics: in the 20C mode, they had a capacity of more than 120 mAh/g after 300 cycles, and in the initial period of cycling (the first 50 cycles), their capacity was about 120 mAh/g even in the 30C mode.

A very interesting composite material is described in [85]. The initial graphene oxide dried in vacuum was thoroughly dispersed in ethylene glycol by ultrasonic treatment, and then an ionic liquid was added to this suspension in which the starting materials for the synthesis of lithium titanate (LiOH, H_2O_2, Ti $(OBu)_4$) were dissolved. After that, a hydrothermal reaction was carried out in an autoclave at a temperature of 180 °C, and then the product was annealed at a temperature of 550 °C in a nitrogen atmosphere. An ionic liquid (octylmethylimidazolinium chloride) served to exfoliate graphene. The electrodes from the described material were characterized by record performances: 162 mAh/g in C/5 mode, 148 mAh/g in 20C mode, and 159 mAh/g after 100 cycles in C/2 mode.

Among the non-carbon components of composites with LTO, designed to provide increased electrical conductivity, titanium nitride, titania, silica and zirconia, as well as metallic silver were mentioned (see e.g., [86, 87]).

The advantages of nanoscale materials for electrodes of lithium-ion batteries are widely recognized. Various explanations are in favor of such materials, of which the first is a reduction in diffusion length as particle sizes decrease. Since the diffusion coefficient of lithium in the solid phase is quite small, it is the diffusion length that determines the capabilities of the fast charges and discharges of lithium-ion batteries. Literature data on the diffusion coefficient of lithium in lithium titanate are scarce and have a significant scatter from 10^{-12} to 10^{-8} cm^2/s.

To directly confirm the advantages of nanoscale titanate over a material with micron particle sizes, the electrochemical characteristics of a material obtained by flame spray pyrolysis, which has characteristic particle sizes of 20 to 30 nm, were compared with the same material annealed at a temperature of 800 °C, and having a characteristic particle size of 1–2 microns [88]. If in C/10 mode both materials showed a specific capacity of about 120 mAh/g, then with a forced discharge (100 C) the capacity of the microsize material dropped to 2 mAh/g, while the capacity of the nanomaterial remained at the level of 70 mAh/g.

The literature describes a wide variety of options for nanoscale lithium titanate, which differ in particle shape and synthesis methods. Particles can have a 3D shape (a simple spherical or close to spherical, the shape of hollow spheres, the shape of various flowers or simply an irregular shape), the 2D form (different plates, sheets) the 1D form (nanofibres, nanowires and nanotubes) or 0D form (nanoparticles).

To synthesize LTO nanoparticles, a diversity of rather classic synthetic methods, such as sol-gel, molten salt, microwave, hydrothermal and solvothermal syntheses have been developed. These methods provide control of stoichiometry, low heat treatment temperature and relatively short synthesis time and can produce nanoparticles with narrow size distributions. It must be stressed that the properties of nanoparticles are strongly dependent on heat treatment temperature.

LTO nanotubes and nanowires are usually prepared by hydrothermal synthesis. Such prepared nanotubes with outer diameter 11 nm and inner diameter as small as 6 nm are described in [89]. The diameter of LTO nanofibers is usually much larger than that of nanowires or nanotubes and may amount up to 300 nm. In spite of this, the electrodes based on nanofibers demonstrate rather high specific capacity even at high rates. For instance, such electrodes showed a capacity of 160 and 138 mAh/g at 0.1C and 10C [90]. It is believed that such good performances are due to the high degree of crystallinity of the nanofibers obtained by electrospinning.

It is worth noting that on going from usual micromaterials to nanomaterials the single-phase ranges of the curve for $Li_4Ti_5O_{12}$ appear to be wider.

The literature on 2D forms of LTO is rather scarce, but published data allow us to consider such objects as very promising. Thin-film LTO electrodes can be prepared by spray pyrolysis, electrostatic spray deposition technique, a pulsed laser deposition technique, ion beam sputtering, a magnetron sputtering and so on. The electrodes based on LTO nanosheets with thickness of 5–7 nm showed specific capacity of 150 mAh/g at low rates and above 100 mAh/g at 60°C. Homogeneous, high-quality crack-free thin films of LTO can also be prepared by spin coating using a rapid heat treatment. Compared with traditional furnace annealing, the rapid heat treatment results in a smaller LTO grain size, higher capacity and more stable cycling behavior.

2.1.4 Miscellaneous materials

In addition to lithium metal, silicon-based materials and lithium titanate, which were already used in commercial lithium-ion batteries, including all-solid-state batteries, a wide variety of negative electrode materials have been investigated and proposed. Other elements of group IV of the periodic table primarily tin and germanium, some oxides, in particular, titanium and tin oxides, as well as nitrides deserve attention.

The tin intermetallic compound with the highest content of lithium has the composition $Li_{4.4}Sn$, which corresponds to theoretical specific capacity of 993 mAh/g. As in the case of silicon, the lithium insertion into tin is accompanied by a very large volume expansion. The natural solution of this problem is using nanomaterials. For instance, it was reported that electrodes based on tin nanowhiskers demonstrated the capacity of 400 mAh/g and sufficiently good cyclability [91]. Anodes of thin tin foil described in [92] had an initial capacity as high as 600 mAh/g but demonstrated quite a fast capacity fading on further cycling. Adequate results were also obtained for films consisting of tin and aluminum [93]. Tin alloys or composites with certain metals proved to be more stable. The Cu_6Sn_5 intermetallic compound has the theoretical specific capacity of 358 mAh/g. The tin alloy with antimony has the theoretical capacity of up to 600 mAh/g. Based on certain alloys, thin-film electrodes were synthesized. For instance, for the film of the Sn/Cu_6Sn_5 composite, the capacity of more than 400 mAh/g was obtained [94]. For Cu_6Sn_5 deposited on a porous copper support, the capacity was about 490 mAh/g. On thin films of the Ni–Sn alloy containing 62 at. % tin, the capacity was 650 mAh/g [95].

Of much greater interest is germanium. Although germanium is inferior to silicon in terms of theoretical gravimetric specific capacity (1600 mAh/g *vs.* 3500 mAh/g), their volumetric specific capacities are rather close (7366 and 8334 mAh/cm³). At the same time, germanium exhibits electrical conductivity two orders of magnitude higher than that of silicon and shows room-temperature Li-ion diffusivity 400 times greater than that of silicon. Like other high-capacity materials, a major disadvantage of germanium is a huge (more than 350%) volume expansion at lithiation. This phenomenon results in cracking and pulverization with a loss of electrical contact. The standard approach to overcome these troubles is to use nanomaterials. Various germanium nanomaterials (nanoparticles, thin films, nanowires, nanotubes, nanofibers, core-shell and yolk-shell structures and so on) are described in literature [96]. Different synthetic techniques have been used in producing Ge-based electrodes, such as 'wet chemistry' method, pyrolysis, dealloying method, Vapor Liquid Solid (VLS) method, Solution Liquid Solid (SLS) method, Chemical Vapor Deposition (CVD) and electrochemical deposition.

For example, [97] describes germanium nanowires synthesized by Solution–Liquid–Solid (SLS) growth using Au nanocrystal seeds from diphenyl germanium (DPG). The electrochemical performances of electrodes with such nanowires strongly depend on the electrolyte nature, and with using 1 M $LiPF_6$ in the mixture of ethylene carbonate and dimethyl carbonate with addition of fluoroethylene carbonate the electrodes demonstrated a capacity of 1300 and 1250 mAh/g after the 5th and 100th cycles.

A similar SLS method was used in [98] for synthesis of germanium nanowires from new oligosilyl germane compounds by using indium nanoparticles as a seeding agent. These nanowires showed specific capacity more than 1000 mAh/g.

Electrodeposition is a very convenient technique for the synthesis of various nanomaterials. A forest of germanium nanowires with a rather high aspect ratio (up to 10^3) was deposited in [99] from a solution of H_2GeO_3 onto a mercury pool cathode. Due to high hydrogen overpotential at Hg the faradaic efficiency of Ge deposition was rather high (ca. 90%). In this case, first a solution of Ge in Hg was formed, and after reaching saturation the nanowires deposition began. A very simple method of germanium nanowires electrodeposition is described in [100]. Here a glass plate with ITO (Indium-Tin-Oxide) coating was used as a cathode. The electrochemical reduction of ITO at the beginning of the process covers the substrate with indium nanoparticles, which act both as a reduction site for dissolved Ge species and as a solvent for crystallization of the Ge nanowires. In [101] Ge nanowires were deposited onto copper foil covered by eutectic GaIn droplets. The negative electrodes based on these nanowires had a specific capacity up to 1350 mAh/g. In [102] Ge nanowires were deposited onto titanium foil covered by arrays of indium nanoparticles. The deposition was carried out from buffer solution of GeO_2 in a potentiostatic mode. Thus the deposited Ge nanowires demonstrated the reversible capacity of lithium insertion-extraction of about 1180 mAh/g, which corresponded to the formation of $Li_{3.05}Ge$ alloy.

As regards to oxide negative electrodes, the most important materials are tin-oxides-based ones. In the first stage, oxides of bi- or tetravalent tin are reduced to form metal nanoparticles which are later insert into lithium. Due to their small size the particles are not destroyed at cycling despite the large volume expansion. The reversible capacity determined only by the lithium insertion/extraction processes is approximately equal to the irreversible capacity diminished to initial oxide reduction. Moreover, in the majority of cases, the capacity of such electrodes turns out to be lower than its theoretical value (300–600 mAh/g) and it is almost halved after the first 10–30 cycles. Such degradation is explained by the gradual recrystallization and coarsening of tin particles.

Thin films of mixed nanostructured tin (90%) and titanium (10%) oxides formed by globules with the size of 10 ± 20 nm, which in turn were built of blocks measuring 6 ± 8 nm, turned out to be a position for negative electrodes of lithium-ion batteries [103]. They were based on a $(Sn,Ti)O_2$ solid solution with the cassiterite structure in the amorphous SnO_2 matrix. These structures could reversibly insert lithium virtually without loss in capacity. Based on the analysis of the lithium reaction products with the electrode material and their electrochemical properties, it was shown that the mechanism of charge/discharge processes in these electrodes is characterized by the involvement of oxygen in the reversible reactions of lithium insertion. The addition of TiO_2 stabilized the oxide structures. Titania is not reduced by lithium and favors the formation of the oxygen environment for the neighboring tin atoms even after their reduction by lithium. The cathodic polarization of such material yields tin metal and $Li_x Sn$ intermetallics. During the subsequent anodic polarization some of the tin is again oxidized, which is stoichiometrically equivalent to the reversible insertion of lithium into the oxide phase.

2.2   Materials for positive electrodes

2.2.1   Lithium cobaltite

The working (discharge) potential of the carbonaceous electrodes used in the first versions of lithium-ion batteries was 0.3–0.6 V more positive than the potential of the lithium electrode. In order for the battery voltage to be sufficiently high, Japanese researchers used lithiated cobalt oxide ($LiCoO_2$, lithium cobaltite) as an active material of the positive electrode. Lithiated cobalt oxide has a potential of about 4.3 V *vs.* lithium electrode, so that the working voltage of the battery has a characteristic value of 3.6–3.7 V. The great merit in creating electrodes based on lithium cobaltite (and, thus, in general in creating lithium-ion batteries) goes to Professor J. Goodenough (USA) [104]. $LiCoO_2$ was the first and the most commercially successful material of the positive electrode, and up to now this material is used in the majority of commercial lithium-ion batteries despite the fact that this material is quite expensive.

When the battery is charged, lithium deintercalation occurs; during discharge, the process proceeds in the opposite direction:

$$LiCoO_2 \rightarrow Li^+ + e + CoO_2, \qquad (2.4)$$

Unfortunately, deep cycling (delithiation above 4.2 V, meaning approximately 50% or more Li extraction) induces lattice distortion from hexagonal to monoclinic symmetry and this change deteriorates the cycling performance (Fig. 2.8). That is why, only *ca.* 50% of lithium is extracted at cycling. The theoretical specific capacity corresponding to Equation (2.4) amounts to 273 mAh/g, whereas real values reach not more than 140 mAh/g. Therefore, for real conditions Equation (2.4) must be rewritten as:

$$LiCoO_2 \rightarrow x\, Li^+ + x\, e + Li_{1-x}CoO_2, \qquad (2.4)$$

with $0 < x < 0.5$.

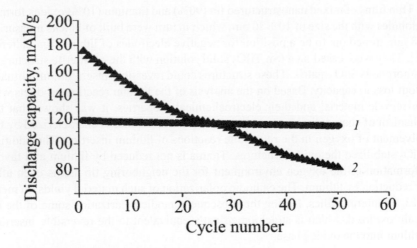

Figure 2.8. The cycling performance of $LiCoO_2$-based electrode with normal and deep delithiation. Charge cut-off voltage 4.1 V (curve 1); 4.4 V (curve 2)

Lithium cobaltite belongs to a family of oxides with a general formula $LiMO_2$, where M denotes cobalt, nickel, chromium or vanadium. All such oxides have a layered structure of rock salt. M^{3+} and Li^+ ions occupy the alternate planes giving a sequence of -O-Li-O-M-O- etc. along the *c* axis (Fig. 2.9). The structure depicted in Fig. 2.9, presents so-called O3 layered structure ('O' refers to octahedral), where there are three MO_2 sheets per unit cell. During charging $LiCoO_2$-based electrode, Li^+ ions are extracted from the crystal structure, their concentration decreases and the concentration of corresponding vacancies increases.

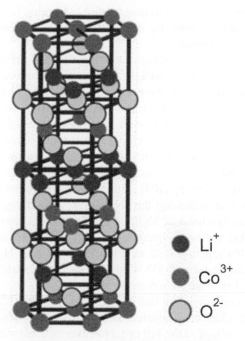

$\bullet$ Li$^+$

$\bullet$ Co^{3+}

$\bigcirc$ O^{2-}

Figure 2.9. Crystal structure of LiCoO$_2$

The synthesis technology of lithiated oxides is based on a variety of high-temperature (sintering, combustion synthesis, molten salt synthesis) and low-temperature (sol-gel, ion exchange, precipitation from solutions) processes. When LiCoO$_2$ is synthesized by conventional sintering in the air at temperatures higher than 800 °C, it acquires the O3 layered structure with high level of ordering of the Li$^+$ and Co^{3+} cations on alternate planes. The starting materials in this synthesis can be thoroughly mixed cobalt carbonate (or cobalt oxide) and lithium carbonate (or lithium hydroxide). At the same time, low-temperature synthesis (<400 °C) results in the formation of spinel-like phase (Li$_2$Co$_2$O$_4$) with huge disordering of the cations. This low-temperature product demonstrates rather poor electrochemical behavior.

Thin-film electrodes from LiCoO$_2$ are applied, in general, by the same methods, that is, radio frequency sputtering [105], sol-gel deposition [106], CVD and pulsed laser deposition [107].

Layered LiCoO$_2$ exhibits a clear anisotropy of transport properties. The ionic conductivity of single-crystal LiCoO$_2$ in different directions varies by several orders of magnitude. At the same time, real LiCoO$_2$ films are, as a rule, polycrystalline, which makes it possible to operate such electrodes with sufficiently high current loads.

It has already been noted above that electrodes based on LiCoO$_2$ can be stably cycled only to a depth of not more than 50%. Irreversible changes with the formation of a monoclinic structure occur when a larger amount of lithium is extracted from LiCoO$_2$. Therefore, further improvements in LiCoO$_2$-based electrodes were aimed precisely at overcoming this drawback. Numerous works on the improvement of

$LiCoO_2$-based electrodes were reduced, mainly, for the improvement of their structure. This is achieved both by different technological methods, and by the introduction of various modifiers.

For example, the crystallographic modification O2 of $LiCoO_2$ was found to have a more pronounced layered structure, and therefore more suitable for reversible lithium insertion than the normal modification O3 [108]. The material with O2 modification could be prepared, in particular, by ion exchange from sodium cobaltite in a solution of lithium bromide in hexyl alcohol. In spite of the O2 material being almost identical to O3 in many respects, it is more stable at cycling. Another technology for the preparation of $LiCoO_2$ with a good texture, i.e., with a predominant orientation of crystallites in the direction of lithium insertion, is based on radio frequency magnetron sputtering [109]. This approach provides the manufacture of thin-film electrodes. For increasing the reversibility of cycling, an original method of supercritical water synthesis of highly dispersed $LiCoO_2$ powders was proposed [110]. Following this method, aqueous solutions of $LiOH$ and $Co(NO_3)_2$ under a pressure of 30 MPa together with an aqueous solution of H_2O_2 under the same pressure and at a temperature of 400 °C are fed into an electric furnace, where the synthesis reaction takes place for a mere 30 seconds.

It has already been noted that one of the advantages of $LiCoO_2$ is a rather high positive potential at which it is lithiated, i.e., positive electrode is discharged. This provides acceptable voltage of the full cell. However, this advantage also turns into a disadvantage, since the charge of the positive electrode occurs at even more positive potentials, at which oxidation of the electrolyte on the catalytically active surface of $LiCoO_2$ or on the carbon additive used to increase the electrical conductivity of the active layer is possible. To overcome this drawback, an original method for modifying the surface of the electrode was proposed [111]. The modification consists in applying to the surface the thinnest (not more than 0.1 μm thick) layer of a diamond-like carbon by plasma spraying from ethylene. The resulting electrodes are stably cycled in the range of potentials from 3.0 to 4.2 V with a specific capacity of more than 200 mAh/g.

Later, a variety of other materials were proposed as coatings of $LiCoO_2$ particles. So, the coating of lithium cobaltite with a SnO_2 film (which is applied by the sol – gel technology followed by annealing) to enhance the stability of the structure during cycling is described in [112]. Among other oxides proposed as such protection coatings ZrO_2, TiO_2, SiO_2, Al_2O_3, and B_2O_3 must be mentioned [113–115]. Figure 2.10 clearly shows the beneficial effect of Al_2O_3 coating on cycling performances of the $LiCoO_2$-based electrode.

In [116], the beneficial effect of a coating of $LiMn_2O_4$, which is applied to $LiCoO_2$ powder, both to protect against electrolyte oxidation on charging and to increase the thermal stability of $LiCoO_2$ itself, is described. (It is known that when heated, $LiCoO_2$ can decompose with oxygen evolution and this process proceeds especially quickly on powders with high dispersion).

Moreover, it turned out that various oxide coatings of $LiCoO_2$ give a possibility of increasing the depth of cycling up to 70% (instead of 50%), which corresponds to specific capacity of 200 mAh/g. [112, 117–121]. In this case charge potential can be increased up to 4.6 V without notable acceleration of capacity fading at the cycling.

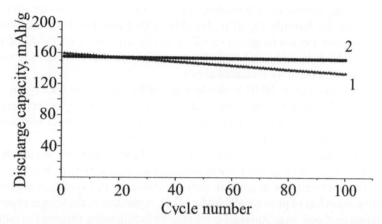

Figure 2.10. The effect of Al_2O_3 coating on cycling performances of $LiCoO_2$-based electrode: 1 – unmodified $LiCoO_2$, 2 – Ai_2O_3-modified $LiCoO_2$

The radical method of increasing the specific capacity (i.e., the depth of cycling), as well as of decreasing the cost is the use of mixed lithiated oxides, i.e., $LiCoO_2$, in which part of the cobalt ions are replaced by ions of one or two other metals. Quite a few of different multicomponent lithiated oxides, of which the most popular were $LiNi_xCo_yMn_zO_2$ and $LiNi_xCo_yAl_zO_2$ systems, and more specifically, $LiNi_{1/3}Co_{1/3}Mn_{1/3}O_2$ (NMC) [122] and $LiNi_{0.8}Co_{0.15}Al_{0.05}O_2$ (NCA) were investigated [123, 124]. Both materials are now considered as environment friendly, quite low-cost products with high specific capacity and good cyclability.

NCA-based positive electrodes are used, in particular, in Panasonic batteries for Tesla electric vehicles. NCA belongs to the family of so-called Nickel-rich Layered Oxides (NLO) [125]. The practical discharge capacity of NCA amounts to ca. 200 mAh/g, and the cyclability of such electrodes is not inferior to that of traditional lithium cobaltite. Certain disadvantages of NCA consist in some instability at elevated temperatures (higher than 50 °C) due to the generation and growth of micro-cracks. To improve the electrochemical and thermal properties of NCA it was suggested to coat its surfaces with metal oxides, including the oxides of aluminum, zinc, zirconium, cerium etc., (see, e.g., [126–128]) as well as with other substances, e. g., with AlF_3 [129], FeF_3 [130], $FePO_4$ [131] and others. A coating of NCA with few-layer reduced graphene oxide is also interesting [132]. The most popular methods of the NCA synthesis are based on co-precipitation process [133, 134] and microwave-assisted approach [131, 135].

NMC having practical specific capacity up to 230 mAh/g (theoretical value is 278 mAh/g), is also widely used in the battery industry. It demonstrates rather good cyclability and thermal stability even at 50 °C. NMC has the same structure as $LiCoO_2$, i.e., it belongs to α-$NaFeO_2$ type of layered rock salt structure. From a formal point of view NMC can be considered as solid solution $LiCoO_2$-$LiNiO_2$-$LiMnO_2$ (1:1:1).

In the initial state, nickel, cobalt and manganese are, respectively, in the 2+, 3+, and 4+ state, and Ni ($2^+/4^+$) and Co ($3^+/4^+$) transitions are realized during cycling, moreover, in the course of delithiation, the Ni^{2+}/Ni^{3+} transition occurs first (with increasing x in the formula $Li_{1-x}[Co_{1/3}Ni_{1/3}Mn_{1/3}]O_2$ from 0 to 1/3), then the Ni^{3+}/Ni^{4+} transition (for x in the range $1/3 < x < 2/3$), and finally, Co^{3+}/Co^{4+} (with increasing x from 2/3 to 1). It is such a scheme of redox processes that provides the value of theoretical specific capacity indicated above.

The stable cycling of NMC is due to a negligible change in the crystal lattice. When 60% of the total lithium contained in $LiNi_{1/3}Mn_{1/3}Co_{1/3}O_2$ is recovered, the volume of a single crystalline cell does not change and amounts to 0.1 nm^3, and with almost complete deletion it decreases to only 0.095 nm^3.

The electrochemical performances of NMC-based electrode materials were found to be strongly dependent on the synthesis process. A variety of NMC synthesis methods are described, but in general, they are similar to the synthesis methods for other active materials of positive electrodes. This, in particular, the sol-gel process, co precipitation methods, hydrothermal synthesis, including using microwave radiation, solid-phase synthesis methods, microemulsion methods, methods of thermal polymerization, methods of 'burning', methods of 'soft chemistry', suspension method (rheological phase), synthesis from melts. In a typical sol-gel process, an aqueous solution of a stoichiometric mixture of lithium, manganese, nickel and cobalt acetates is prepared to which citric acid is added as a chelating agent. After adjusting the pH, the solution is evaporated to obtain a clear gel. This gel is dried in air at a temperature of 120 °C and then annealed at a temperature of 450 °C until the organic components are completely carbonized. The resulting product is ground in a ball mill and sintered at a temperature of 850 °C.

The coprecipitation method typically uses the precipitation of a mixture of hydroxides. In particular, when using a solution of a mixture of nickel, cobalt and manganese acetates, the precipitation is carried out with a solution of a mixture of lithium and ammonium hydroxides with continuous stirring. After separation and washing of the precipitate, it is dried at a temperature of 120 °C, pelletized, annealed at a temperature of 400 °C and again milled and pelletized. These tablets are sintered at a temperature of 900 °C.

Solid phase synthesis usually also involves lithium, nickel, manganese and cobalt acetates. A stoichiometric mixture of these components with the addition of excess oxalic acid is processed in a ball mill, then dried at a temperature of 120 °C and annealed immediately.

In the suspension NMC synthesis method used in [136], the starting materials were cobalt and manganese oxides and nickel and lithium hydroxides. A mixture of these substances with a stoichiometric excess of lithium hydroxide was thoroughly ground, a small amount of water was added and autoclaved at a temperature of 90 °C. The resulting 'rheological phase' was fired in air at a temperature of 800 °C for 20 hours. In the synthesis from melts the authors of [137] proceeded from a solution of nickel, cobalt, and manganese nitrates in the eutectic melt of lithium nitrate and chloride. This initial melt was kept at a given temperature (in the range from 750 to 950 °C), and after cooling it was crushed, washed repeatedly with water and dried at a temperature of 150 °C.

The electrochemical performance of NMC-based electrode materials could be also improved by doping cations (Al^{3+}, Fe^{3+}, Mg^{2+}) and anion substitution (F^- for

O^{2-}). For instance, it was shown that substitution of a bit of Al^{3+} ions for Ni^{2+} up to composition $LiNi_{0.32}Mn_{0.33}Co_{0.33}Al_{0.01}O_2$ without changing the phase composition and particle morphology resulted in increase of cycling stability, i.e., in diminishing capacity fading at cycling. This effect was explained by the increase of Li^+ diffusion coefficient due to decrease of the cation mixing in Li layer. Al^{3+} substitution for Co^{2+} improves the thermal stability of NMC even at substitution level as low as 0.06.

2.2.2 Lithium iron phosphate

Lithium iron phosphate $LiFePO_4$ is a fundamentally different type of active material for positive electrodes of lithium-ion batteries, including all-solid-state thin-film ones. This material belongs to the family of polyanionic materials. In this case, the large trivalent anion PO_4^{3+} stabilizes the crystalline structure. The first mention of the electrochemical activity of $LiFePO_4$ was in 1997 [138]. In the same work, a description was given of the processes occurring in the material during the extraction and reverse insertion of lithium, and also the main problems with its use in the battery were noted. The huge interest in this material is associated with its following advantages: (1) a high degree of stability of $LiFePO_4$ in the olivine structure, which allows it to withstand a large number of charge-discharge cycles without significant capacity fading, (2) the ability to work in extreme conditions, in particular at high temperatures, (3) the ability to achieve a capacity close to the theoretical value of 170 mAh/g, even at high current loads. However, along with the listed advantages, $LiFePO_4$ is characterized with disadvantages, such as low electronic conductivity and low diffusion coefficient of lithium, which significantly limits the depth of the process in the volume of $LiFePO_4$ particles. Numerous works to eliminate these undesirable phenomena are reduced to three main areas: (1) the creation of an electrically conductive coating on the surface of $LiFePO_4$ particles, (2) doping of $LiFePO_4$ with cations of other metals or with F, Cl^- anions, (3) reduction of the particle size of the material, i.e., reducing the diffusion length.

Immediately after the publication of Ref. [138], a proposal was made to apply a thin coating of carbon to each $LiFePO_4$ particle. In addition, in the manufacture of the active mass of the electrode, along with $LiFePO_4$ (regardless of the coating of its particles with carbon), a certain amount of carbon black is introduced into it, as a conductive additive, which promotes the distribution of the electrochemical process throughout the entire depth of the electrode. The set of methods for improving the electrochemical characteristics of the electrode by introducing carbon additives into the active mass is quite extensive: joint grinding with carbon black, sucrose coating (followed by heat treatment), mechanical activation with a carbon-containing component, and the use of carbon gel [139–142]. The main methods of introducing carbon into the electrode material per se include: the addition of carbon black to precursors during the synthesis [143–146], the sol–gel method [147–149], a method based on the formation of carbon during the combustion of organic additives, including citric acid [150], glycine [151], urea [152], polypyrrole [153] and various carboxylic acids [154, 155]. It is important to emphasize that the carbon coating on olivine particles not only increases the electronic conductivity of the active mass, but also prevents the formation of ferric iron and prevents the agglomeration of particles of the active substance.

A large number of studies were devoted to doping, i.e., the replacement of some iron or lithium atoms with cations of various metals (Ti, Al, Ca, V, Mg, Cu, Zn, Ru, Mn, Ti) to increase the electronic conductivity of the $LiFePO_4$ electrode [156–171]. Doping of olivine is carried out by a small amount of metal ions having a close ionic radius with a replaced ion. In this case, there is no distortion of the structure and changes in the spatial arrangement of atoms. $LiFePO_4$ can be doped with both isovalent (divalent) and heterovalent ions. It is assumed that cations of preferably a small radius occupy the crystallographic positions of lithium. For example, Mg^{2+} ions, due to their small radius (smaller than the radius of Fe^{2+}), replace lithium ions, creating a cation-deficient compound $Li_{0.98}Mg_{0.02}FePO_4$, whose reversible capacity is 160 mAh/g [168]. It was noted in [169] that doping provides the ability to withstand large currents at fast charges/discharges. This ability of the material is explained by the fact that substitution of a part of Fe^{2+} with Mg^{2+} weakens the $Li - O$ bond, which, in turn, leads to a high mobility of lithium ions and a high diffusion coefficient of lithium.

Like lithium titanate described above, lithium iron phosphate forms a two-phase system with very narrow regions of existence of solid solutions. In general, one can assume that during the oxidation of $LiFePO_4$, i.e., when the positive electrode is charged, the $FePO_4$ phase is formed, and the entire charge process reduces to a change in the relative amount of these phases, almost without changing the electrode potential. Such 'flat' charging and discharge curves represent a definite advantage of this material.

2.2.3 Vanadium oxides

The principal advantage of vanadium oxides over all traditional positive electrode materials, including the lithium cobaltite and lithium iron phosphate described above, is that the insertion of lithium into traditional oxides is limited by change of no more than one unit valency of the oxide-forming metal, whereas during reduction, for example, vanadium pentoxide, vanadium valency can, in principle, change by three units (from +5 to +2). Thus, it can theoretically be expected that vanadium oxides will have a much higher specific capacity. Indeed, a hypothetical process

$$V_2O_5 + 6\,Li + + 6\,e \rightarrow Li_6V_2O_5 \qquad (2.5)$$

according to the laws of Faraday corresponds to a discharge capacity of 883.5 mAh/g, that is 5–6 times higher than the capacity of traditional materials of the positive electrode of commercial lithium-ion batteries.

Vanadium oxides themselves represent the charged form of the positive electrode material. When assembling the battery, they must be combined with the charged form of the negative electrode material, for example, with lithium metal, which significantly limits the applicability of such materials.

Vanadium oxides have been widely used in primary chemical power sources with a lithium anode. Unfortunately, the insertion of lithium into the crystal lattice of vanadium oxide is associated with significant structural changes. Layered structure V_2O_5 (α-V_2O_5 phase) is characterized by a weak bond between the layers, which facilitates reversible insertion of Li^+ cations. In the range of potentials from 3.5 to

2.5 V, a reversible insertion of 1 mole of lithium per 1 mole of V_2O_5 occurs with the formation of the δ-LiV_2O_5 phase, which corresponds to specific capacity of 147.2 mAh/g. Already at the insertion of 2 moles of lithium per mole of V_2O_5 a γ-$Li_2V_2O_5$ phase appears with a irreversible structural change. Unlike lithium iron phosphate materials based on vanadium oxides operate in a fairly wide range of potentials. The interval of working potentials of electrodes based on vanadium oxides significantly depends on the oxide structure even for the same stoichiometric composition.

Recently, there have been many works where the possibility of reversible lithium insertion into vanadium pentoxide by using electrodes in the form of very thin films or when using nanostructures were shown. Thin film vanadium oxide-based electrodes are mainly intended for micro batteries, including all-solid-state ones. Many methods for the manufacture of thin-film electrodes from vanadium oxides have been described, including radio frequency magnetron sputtering V_2O_5 [172, 173], pulsed laser deposition [174], ion beam deposition [175], chemical vacuum deposition of amorphous V_2O_5 [176, 177], plasma-assisted vacuum deposition [178] anodic or thermal oxidation of a metal. In [179] a thin layer of vanadium metal was first sprayed onto a substrate of stainless steel and thus, these samples were oxidized in air at a temperature of 800 K for some time. An increase in oxidation time leads to deeper oxidation and to increase the capacity.

In the reference [180], thin-film electrodes prepared by deposition of V_2O_5 xerogels onto the rotating substrate, and in [181]—prepared by heat treatment of V_2O_5 gel are described. On thin-film electrodes, rather high values of specific capacity were reported. For example, in [177] for films with a 200 nm thickness, capacities of more than 450 mAh/g were obtained for 100 cycles at a current C/10.

As for nanostructures, the 1D forms of vanadium oxides are the most popular. V_2O_5 nanofibers [182], nanorods [183], nanotubes [184, 185], nanowires [186], nanobelts [187–189] must be mentioned. Nanoribbons of partially lithiated vanadium pentoxide are characterized by the maximum achieved by the present capacity for the insertion and extraction of lithium. Such objects with a thickness of 10–15 nm, a width of 200–300 nm and a length of 10 microns showed an initial capacity of about 500 mAh/g with acceptable stability during cycling.

Of the materials based on pentavalent vanadium oxide, a special interest has been given to lithium trivanadate LiV_3O_8 with a layered structure that consists of ribbons composed of [VO_6] octahedra and trigonal prisms [VO_5]. These ribbons are interconnected because of Li^+ ions. Such a structure provides the possibility of the reversible insertion of a relatively large amount of lithium, at least 4.5 Li^+ ions per formula unit LiV_3O_8, which corresponds to specific capacity of 419 mAh/g. The amorphous LiV_3O_8 has notable advantages compared with crystalline one [190]. Lithium trivanadate is also often used as thin films or in the form of nanostructured materials.

2.3 Materials for solid electrolytes

It has already been noted above that in all-solid-state thin-film batteries, electrodes can be made, in principle, from the same materials as in traditional lithium-ion batteries. A significant exception is lithium metal, the possibility of using this represents a significant advantage of solid-state batteries. The electrolyte of all-solid-state thin-film batteries is fundamentally different from the electrolyte of traditional lithium-ion batteries, and many studies have been devoted to the development and improvement of such electrolytes [191–203].

The electrolytes for all-solid-state batteries should have high ionic conductivity and low electronic conductivity at an operating temperature (preferably ambient temperature), a wide window of electrochemical stability, manufacturability and be compatible with electrodes. The latter feature suggests that the electrolyte must be stable against chemical reaction with the electrodes, especially with Li metal or Li alloy negative electrodes and have matching thermal expansion coefficients with both electrodes. Both crystalline and amorphous materials are used as solid electrolytes. LiPON is a typical representative of amorphous (glassy) electrolyte for all-solid-state thin-film batteries. Other examples of amorphous solid electrolytes are oxide and sulfide glasses. The crystalline solid electrolytes are represented by A-site deficient perovskite solid solutions, NASICON-type Li ion conductors, LISICON and thio-LISICON-type Li ion conductors and Garnet-type Li ion conductors.

2.3.1 Amorphous solid electrolytes

LiPON was already described in detail in Chapter 1. LiPON has a phosphate-derived structure, in which nitrogen atoms substitute oxygen bridging atoms of the $-PO_4$ groups in the glassy network. The presence of these nitrogen atoms sufficiently increases the chemical stability of phosphate, as well as the electrolyte hardness and the glass-transition temperature. The salient feature of LiPON consists in its negligible reactivity with the lithium. The high ionic conductivity of LiPON is related to N doping, probably due to the formation of cross-linked NP3 structures. It was mentioned in the Chapter 1 that LiPON was first obtained by the team of Bates through the means of a RF magnetron sputtering process, by using a Li_3PO_4 target in a nitrogen reactive atmosphere. Later, many other attempts were successively carried out to deposit thin films of LiPON with other physical techniques, including pulse laser deposition [204, 205], electron beam evaporation, ion beam processes [206]. However, the magnetron sputtering appeared to be the best compromise in terms of versatility, economic impact and film quality. It must be stressed that the transport properties of the LiPON thin films sufficiently depend on the deposition conditions. Quite high conductivity value measured was $3.3*10^{-6}$ S/cm at room temperature for the films with composition of $Li_{2.9}PO_{3.3}N_{0.46}$ [207] deposited by means of RF magnetron sputtering by using a Li_3PO_4 target in N_2/O_2 reactive atmosphere. Overall, the increase of nitrogen partial pressure, target density and sputtering power seems to enhance the ionic conductivity of the films [208–210].

Another interesting example of LiPON-like glassy electrolyte is the so-called LiSON with typical composition $Li_{0.29}S_{0.28}O_{0.35}N_{0.09}$ and an ionic conductivity $2*10^{-5}$ S/cm [211]. This material was manufactured by RF magnetron sputtering by using Li_2SO_4 target in a pure nitrogen atmosphere. Yet another similar glassy electrolyte, known as LiPOS ($6LiI–4Li_3PO_4–P_2S_5$) also possesses acceptable conductivity [212].

Sulfide glassy solid electrolytes, especially glasses with high Li^+ ion concentration possess, on the whole higher conductivity than LiPON-like electrolytes. In the system $Li_2S-P_2S_5$, provided Li_2S content to be more than 70 mol%, the electrolytes have a conductivity of over 10^{-4} S/cm, one-and-half orders of magnitude higher than conductivity of LiPON-like counterparts. At the same time, it is difficult to synthesize sulfide glasses with rather high Li^+ ion concentration because of easy crystallization during the cooling process. That is why such glasses are prepared by twin-roller rapid-quenching or mechanical milling techniques. The combination of two anions is an effective way of increasing conductivity of glasses (so-called 'mixed-anion effect'). The addition of lithium salts such halides, borohydride or orthophosphates result in enhancing conductivity of glasses up to 10^{-3} S/cm at room temperature. Table 2.1 presents some data on conductivity of sulfide glassy solid electrolytes.

Table 2.1. Conductivity of sulfide glassy solid electrolytes at room temperature

Composition	Conductivity, µS/cm	Reference
$75Li_2S-25P_2S_5$	0.2	213
$67Li_2S-33P_2S_5$	0.1	214
$70Li_2S-30P_2S_5$	0.054	215
$75Li_2S-21P_2S_5-4P_2O_5$	no less than 0.1	216
$56Li_2S-24P_2S_5-20Li_2O$	no less than 0.1	217
$67.5Li_2S-25P_2S_5-7.5Li_2O$	0.11	218
$80(70Li_2S-30P_2S_5) - 20LiI$	0.56	219
$95(80Li_2S-20P_2S_3) - 5LiI$	2.7	220
$67(75Li_2S-25P_2S_5) -33LiBH_4$	1.6	221

Purely oxide glassy solid electrolytes (besides LiPON) are of less interest due to their low conductivity. For instance, the conductivity of $50Li_4SiO_4-50Li_3BO_3$ (or $7Li_2O-2SiO_2-B_2O_3$) is as low as 4 µS/cm at room temperature. The conductivity of $Li_2O-SiO_2-Al_2O_3$ amounts to 3 µS/cm only.

Intermediate forms, the so-called glass-ceramic electrolytes, represent a certain palliative. They have higher conductivity than pure amorphous but a lower one than crystalline electrolytes. Such glass-ceramic electrolytes could be prepared by crystallization of true glass electrolytes. Precipitation of thermodynamically stable crystalline phases from a precursor glass results in reducing grain-boundary resistance. For instance, [222, 223] describe glass-ceramic electrolytes obtained by heat-treating of $Li_2O-Al_2O_3-TiO_2-P_2O_5$ and $Li_2O-Al_2O_3-GeO_2-P_2O_5$ glasses. The mother glasses were prepared using a conventional melt-quenching method. Heat-treatments were performed at 700–1000 ºC, i.e., at the temperatures above crystallization temperatures for 12 hours. The crystalline phases in the above systems had the compositions $LiTi_2(PO_4)_3$ and $LiGe_2(PO_4)_3$, conductivity depending on contents of these phases. Maximal conductivity 1.3 mS/cm was achieved in the system $Li_2O-Al_2O_3-TiO_2-P_2O_5$ heat-treated at 950 ºC. In the system $Li_2O-Al_2O_3-GeO_2-P_2O_5$ maximal conductivity 0.4 mS/cm corresponded to brutto-composition $Li_{1.6}Al_{0.6}Ge_{1.4}(PO_4)_3$.

Yet higher conductivity is inherent to glass-ceramics $70Li_2S-30P_2S_5$ [215], $80Li_2S-20P_2S_5$ [213], $Li_{3.25}P_{0.95}S_4$ [225] and $Li_7P_3S_{11}$ [226]. Glass-ceramics

$70Li_2S-30P_2S_5$ was synthesized by heat-treatment of corresponding glass at the temperature of about 240 °C, barely higher than the crystallization temperature. This treatment resulted in increase of conductivity at room temperature up to 3.2 mS/cm. The conductivity of glass-ceramics $80Li_2S-20P_2S_5$ amounts to 0.74 mS/cm. Glass-ceramics $Li_{3.25}P_{0.95}S_4$ and $Li_7P_3S_{11}$ demonstrate room-temperature-conductivity of 1.3 and 17 mS/cm. Glass-ceramic electrolyte $Li_7P_3S_{11}$ $(70Li_2S-30P_2S_5)$ possessed not only the highest conductivity, but the lowest conduction activation energy of 17 kJ/mol at room temperature (and therefore, the weakest conduction temperature dependence). The latter electrolyte was prepared by heat-treatment of the corresponding glass after grinding it into a powder. Then this powder was densified by pressing at 94 MPa and heated at 280–300 °C for 2 hours.

Unfortunately, $Li_2S-P_2S_5$ glasses and glass–ceramics have a rather low chemical stability in air because they undergo hydrolysis by water vapors. The chemical stability of the sulfide glasses and glass-ceramics can be enhanced by partial substitution of oxides (Li_2O or P_2O_5) for corresponding sulfides (Li_2S or P_2S_5). At the same time the stability of the sulfide glasses and glass-ceramics in dry air is quite satisfactory.

2.3.2   Crystalline solid electrolytes

A-site-deficient perovskite solid solutions with Li ion conductivity (Li-ADPESSs) based on titanates with common formula $Li_{3x}La_{2/3-x}\square_{1/3-2x}TiO_3$, (where square denotes lattice vacancy, often this symbol is omitted in the formula) are the most popular solid electrolytes of the family of the perovskite-type (ABO_3) oxides, with A = Li, La and B = Ti [226, 227]. Usually $0.04<x<0.17$, in this case the abbreviation LLTO is used. Such electrolytes have room-temperature conductivity of about 1 mS/cm. Li-ion conductivity of LLTO strongly depends on lithium content, i.e., on index x, being maximal at x = 0.11 (1.1 mS/cm). At temperatures below 125 °C, the temperature dependence of conductivity follows the Arrhenius law with activation energy 0.37 eV. At higher temperatures LLTO conductivity can be fitted to a Vogel–Tammann–Fulcher relationship, which might involve tilting of TiO_6 octahedra.

The high conductivity of the LLTO is due to the features of their crystal structure, namely, a high concentration of vacancies. It is the vacancy mechanism that provides the motion of lithium ions through square planar bottleneck between A sites, formed by four O^{2-} ions between two neighboring A sites [228].

Li-ADPESSs can be manufactured by various synthetic techniques including sol-gel synthesis, solid-state reactions and so-called floating zone reactions. Sol-gel synthesis is the best for preparing thin-film electrolyte for all-solid-state lithium-ion batteries.

Unfortunately the presence of the easily reducible Ti^{4+} ions hinders Li-ADPESSs use as electrolyte materials in batteries with direct contact with lithium metal anode. Partial reduction of Ti^{4+} to Ti^{3+} on direct contact with lithium metal results in the appearance of electronic conductivity. This problem could be solved with limited substitution of La- or Ti-ions by foreign cations. It is worth noting that such a substitution (doping) in certain cases results in some conductivity increase. Another way to eliminate the electronic conductivity consists in synthesis of Li-

ADPESS in amorphous form [229]. This approach provided a possibility to make an all-solid-state microbattery of the LiCoO$_2$/LLTO/Li system, which withstood 300 charge-discharge cycles.

The canonic NASICON crystallographic structure is NaA$^{IV}_2$(PO$_4$)$_3$, where A^{IV}=Ge, Ti and Zr. The structure can be visualized as a skeleton [A$_2$P$_3$O$_{12}$]$^-$ constituted of AO$_6$ octahedra and PO$_4$ tetrahedra. These polyhedra form interconnected channels and two types of interstitial positions (M' and M'') where conductor cations can be distributed. These cations move from one site to another through bottlenecks, the size of which depends on the nature of the skeleton ions and on the carrier concentration in both type of sites (M' and M''). The most popular Li$^+$-conducting electrolyte with NASICON-like structure is Li$_{1.3}$Al$_{0.3}$Ti$_{1.7}$(PO$_4$)$_3$ (LATP), belonging to the family with common formula Li$_{1+x}$Ti$_{2-x}$M$_x$(PO$_4$)$_3$ (M=Al, Ga, In, Sc). Among Li$^+$-conducting electrolyte with NASICON-like structure, the highest room temperature conductivity of 3 mS/cm has Li$_{1+x}$Al$_x$Ge$_{2-x}$(PO$_4$)$_3$ (LAGP), while zirconium-based counterpart, LiTi$_{0.5}$Zr$_{1.5}$(PO$_4$)$_3$ has lower conductivity. Silicon-substituted electrolyte, in which a part of phosphorus is replaced by silicon Li$_{1+x+y}$Ti$_{2-x}$Al$_x$Si$_y$(PO$_4$)$_{3-y}$ is quite interesting.

LISICON is a family of a solid crystalline electrolyte with a general formula Li$_{2+2x}$Zn$_{1-x}$GeO$_4$. The first representative of such a structure, Li$_{14}$ZnGe$_4$O$_{16}$ was described as early as in 1977. Unfortunately, the room-temperature ionic conductivity of LISICON per se is relatively low, about 10^{-6} S/cm. That is why, LISICON is of interest only for some high-temperature applications. Moreover, LISICON is highly reactive with lithium metal and atmospheric CO$_2$.

Great advantages over LISICON have the electrolytes belonging to the category of thio-LISICON [230]. Among them, crystalline Li$_{3.25}$Ge$_{0.25}$P$_{0.75}$S$_4$ (which can be considered as Li$_{4-x}$Ge$_{1-x}$P$_x$S with x=0.75) has the highest conductivity 2.2 mS/cm, which is comparable with that of Li$_2$S–P$_2$S$_5$ glass–ceramics. At the same time, Li$_2$S–GeS$_2$–P$_2$S$_5$ electrolytes were found to be incompatible with a graphite negative electrode. To overcome this problem the authors of [231] proposed a battery design with a two-layer solid electrolyte. The layer facing the negative (graphite) electrode is LiI–Li$_2$S–P$_2$S$_5$ glass, whereas the layer facing the positive (LiCoO$_2$) electrode is Li$_2$S–GeS$_2$–P$_2$S$_5$ crystalline material. The former electrolyte is known to be stable to electrochemical reduction, and the latter electrolyte is stable to oxidation.

At the same time, thio-LISICON is quite stable in a contact with lithium or lithium-alloy electrodes and has a rather wide electrochemical window. Reference [232] describes the all-solid-state sheet lithium-ion battery with thio-LISICON electrolyte, Li–Al negative electrode, and Mo$_6$S$_8$ as a material of positive electrode. Such a laboratory battery showed good charge–discharge characteristics without capacity fading at a current density of 0.011 mA/cm^2.

It has been found that a similar superionic conductor, namely Li$_{10}$GeP$_2$S$_{12}$ with a special crystal structure demonstrates a higher conductivity of 12 mS/cm [233]. It is worth noting that the conductivity of Li$_{10}$GeP$_2$S$_{12}$ exceeds that of usual liquid organic electrolytes. This new solid electrolyte has many advantages including stability, safety and attractive electrochemical properties (high conductivity and wide potential window).

Of late, solid electrolytes with garnet-related structures have gained a great deal of attention. The ideal garnets can be presented by a general formula of $A_3B_2(XO_4)_3$ where A = Ca, Mg, Y, La or rare earth; B = Al, Fe, Ga, Ge, Mn, Ni or V; X = Si, Ge or Al. It must be stressed that A, B and C are eight, six and four oxygen coordinated cation sites, which crystallize in a face centered cubic structure. In the typical structure of garnets $(A_3B_2X_3O_{12})$ XO_4 tetrahedra and BO_6 octahedra are connected by edge sharing. The most important feature of a garnet structure is its ability to insert Li^+ ions in the structure. The garnets typically contain five to seven Li atoms per formula unit and are referred to as Li-rich ('Li-stuffed') garnets. Increasing the number of lithium atoms per formula unit to five, such as in $Li_5La_3B'_2O_{12}$ (B' = Bi, Sb, Na, Ta), results in a three orders of magnitude increase in ionic conductivity up to $2*10^{-5}$ S/cm [234]. Partial substituting of Zr in Li-stuffed garnet $Li_7La_3Zr_2O_{12}$ by Nb gives a possibility to obtain the material with lithium ion conductivity as high as 0.8 mS/cm [235]. $Li_7La_3Zr_2O_{12}$ doped by Ga has ionic conductivity of 0.54 mS/cm [236]. For substituted garnet $Li_{6.75}La_3Zr_{1.75}Ta_{0.25}O_{12}$ the room temperature conductivity of 0.9 mS/cm was reported [23]. The doping of garnet-structured electrolyte with bromide anion results in 2–3-fold increase of Li^+-ion conductivity.

Initially, $Li_7La_3Zr_2O_{12}$ compounds with a garnet structure were synthesized by solid-state reaction at the temperature as high as 1273 °C [238]. Later, repeated attempts were made to lower the synthesis temperature, including the realization of sol-gel synthesis, but in any case, the technology for producing garnet included high-temperature sintering (not lower than 700 °C) [239, 240]. It is worth noting that the high sintering temperature results in a coarsening of the microstructure and an increase in the grain boundary conductivity. At the same time, the high sintering temperature causes lithium deficiencies in the structure; so extra lithium is required to compensate for lithium losses during high temperature sintering.

By introducing such a dopant as Al can stabilize the cubic $Li_7La_3Zr_2O_{12}$ at lower temperatures, thereby preventing lithium loss that occurs as a result of high temperature sintering. The Al dopant provides two main improvements in the formation of the cubic $Li_7La_3Zr_2O_{12}$. First, the Al atom acts as a sintering aid to lower the heat treatment temperature. Second, Al atoms substitute in both the tetrahedral and octahedral Li sites, which increase the Li vacancy concentration, facilitating Li transport [241].

The most radical way to avoid the complications associated with high-temperature sintering (in particular, lithium loss) is transition to pulsed laser annealing [196]. The effective heating time in pulsed laser annealing is in the order of nanoseconds as opposed to furnace sintering (more than 10^4 s). The rapid heating and cooling do not provide sufficient time for Li atoms to escape from the sample by diffusion. Very high local temperatures as well as incredibly fast heating and quenching rates are features of pulsed laser annealing; this makes lasers ideal for annealing $Li_7La_3Zr_2O_{12}$. The typical parameters of pulsed laser annealing are the pulse duration 25 ns, a pulse repetition rate about 1 Hz, the energy fluence from 100 to 700 mJ/cm². The size of the laser beam must be slightly larger than the sample size so that the entire sample area was irradiated under the same conditions.

References

1. Bates, J.B., Dudney, N.J., Gruzalski, G.R., Zuhr, R.A., Choudhury, A., Luck, C.F. 1992. Electrical properties of amorphous lithium electrolyte thin films. *Solid State Ionics* 53–56: 647–654.
2. Dudney, N. 2008. Thin film micro-batteries. *Interface* No. 3: 44–8.
3. Bates, J.B., Dudney, N.J., Neudecker, B., Ueda, A., Evans, C.D. 2000. Thin-film lithium and lithium-ion batteries. *Solid State Ionics* 135: 33–45.
4. Neudecker, B.J., Dudney, N.J., Bates, J.B. 2000. "Lithium-Free" thin-film battery with *in situ* plated Li anode. *J. Electrochem. Soc.* 147: 517–23.
5. Liu, W.-Y., Fu, Z.-W., Qin, Q.-Z. 2008. A "Lithium-Free" thin-film battery with an unexpected cathode layer. *J. Electrochem. Soc.* 155: A8–A13.
6. Zhou, Y.N., Xue, M.Z., Fu, Z.-W. 2013. Nanostructured thin film electrodes for lithium storage and all-solid-state thin-film lithium batteries. *J. Power Sources* 234: 310–32.
7. Zuo, X., Zhu, J., Müller-Buschbaum, P., Cheng, Y.-J. 2017. Silicon based lithium-ion battery anodes: A chronicle perspective review. *Nano Energy* 31: 113–43.
8. Feng, K., Li, M., Liu, W., Kashkooli, A.G., Xiao, X., Cai, M., Chen, Z. 2018. Silicon-based anodes for lithium-ion batteries: From fundamentals to practical applications. *Small* 14: Article No. 1702737.
9. Zhang, W.-j. 2011. A review of the electrochemical performance of alloy anodes for lithium-ion batteries. *J. Power Sources* 196: 13–24.
10. Szczech, J.R., Jin, S. 2011. Nanostructured silicon for high capacity lithium battery anodes. *Energy Environ. Sci.* 4: 56–72.
11. Liang, B., Liu, Y., Xu, Y. 2014. Silicon-based materials as high capacity anodes for next generation lithium ion batteries. *J. Power Sources* 267: 469–90.
12. Bourderau, S., Brousse, T., Schleich, D.M. 1999. Amorphous silicon as a possible anode material for Li-ion batteries. *J. Power Sources* 81-82: 233–6.
13. Ohara, S., Suzuki, J., Sekine, K., Takamura, T. 2003. Li insertion/extraction reaction at a Si film evaporated on a Ni foil. *J. Power Sources* 119-121: 591–6.
14. Jung, H., Park, M., Yoon, Y.-G., Kim, G.-B., Joo, S.-K. 2003. Amorphous silicon anode for lithium-ion rechargeable batteries. *J. Power Sources* 115: 346–51.
15. Takamura, T., Ohara, S., Uehara, M., Suzuki, J., Sekine, K. 2004. A vacuum-deposited Si film having a Li extraction capacity over 2000 mAh/g with a long cycle life. *J. Power Sources* 129: 96–100.
16. Ohara, S., Suzuki, J., Sekine, K., Takamura, T. 2004. A thin film silicon anode for Li-ion batteries having a very large specific capacity and long cycle life. *J. Power Sources* 136: 303–6.
17. Uehara, M., Suzuki, J., Tamura, K., Sekine, K., Takamura, T. 2005. Thick vacuum deposited silicon films suitable for the anode of Li-ion battery. *J. Power Sources* 146: 441–4.
18. Salah, M., Murphy, P., Hall, C., Francis, C., Kerr, R., Fabretto, M. 2019. Pure silicon thin-film anodes for lithium-ion batteries: A review. *J. Power Sources* 414: 48–67.
19. Mukanova, A., Jetybayeva, A., Myung, S.-T., Kim, S.-S., Bakenov, Z. 2018. A mini-review on the development of Si-based thin film anodes for Li-ion batteries. *Materials Today Energy* 9: 49–66.
20. Takamura, T., Uehara, M., Suzuki, J., Sekine, K., Tamura, K. 2006. High capacity and long cycle life silicon anode for Li-ion battery. *J. Power Sources* 158: 1401–4.
21. Deng, H.X., Chung, C.Y., Xie, Y.T., Chu, P.K., Wong, K.W., Zhang, Y., Tang, Z.K. 2007. Improvement of electrochemical performance of Si thin film anode by rare-earth La PIII technique. *Surf. Coating. Technol.* 201: 6785–8.

22. Cui, L.-F., Hu, L., Choi, J.W., Cui, Y. 2010. Light-weight free-standing carbon nanotube-silicon films for anodes of lithium ion batteries. *ACS Nano* 4: 3671–8.

23. Lu, Z., Zhu, J., Sim, D., Zhou, W., Shi, W., Hng, H.H., Yan, Q. 2011. Synthesis of ultrathin silicon nanosheets by using graphene oxide as template. *Chem. Mater.* 23: 5293–5.

24. Emets, V.V., Kulova, T.L., Skundin. A.M. 2017. Dynamic behavior of silicon-based electrodes at open circuit conditions. *Int. J. Electrochem. Sci.* 12: 2754–62.

25. Arie, A.A., Song, J.O., Lee, J.K. 2009. Structural and electrochemical properties of fullerene-coated silicon thin film as anode materials for lithium secondary batteries. *Mater. Chem. Phys.* 113: 249–54.

26. Roginskaya, Yu.E., Kulova, T.L., Skundin, A.M., Bruk, M.A., Klochikhina, A.V., Kozlova, N.V., Kal'nov, V.A., Loginov, B.A. 2008. The structure and properties of a new type of nanostructured composite Si/C electrodes for lithium ion accumulators. *Russ. J. Phys. Chem. A* 82: 1655–62.

27. Roginskaya, Yu.E., Kulova, T.L., Skundin, A.M., Bruk, M.A., Zhikharev, E.N., Kal'nov, V.A., Loginov, V.B. 2008. New Type of the nanostructured composite Si/C electrodes. *Russ. J. Electrochem.* 44: 1197–1203.

28. Li, W., Yang, R., Wang, X., Wang, T., Zheng, J., Li, X.J. 2013. Intercalated Si/C films as the anode for Li-ion batteries with near theoretical stable capacity prepared by dual plasma deposition. *J. Power Sources* 221: 242–6.

29. Kim, J.-B., Lim, S.-H., Lee, S.-M. 2006. Structural change in Si phase of Fe/Si multilayer thin-film anodes during Li insertion/extraction reaction. *J. Electrochem. Soc.* 153: A455–A458.

30. Hwang, C.-M., Park, J.-W. 2011. Electrochemical characterizations of multi-layer and composite silicon–germanium anodes for Li-ion batteries using magnetron sputtering. *J. Power Sources* 196: 6772–80.

31. Demirkan, M.T., Trahey, L., Karabacak, T. 2015. Cycling performance of density modulated multilayer silicon thin film anodes in Li-ion batteries. *J. Power Sources* 273: 52–61.

32. Kulova, T.L., Mironenko, A.A., Skundin, A.M., Rudy, A.S., Naumov, V.V., Pukhov, D.E. 2016. Study of silicon composite for negative electrode of lithium-ion battery. *Int. J. Electrochem. Sci.* 11: 1370–81.

33. Schmidt, V., Wittemann, J.V., Gösele, U. 2010. Growth, thermodynamics, and electrical properties of silicon nanowires. *Chem. Rev.* 110: 361–88.

34. Zamfir, M.R., Nguyen, H.T., Moyen, E., Lee, Y.H., Pribat, D. 2013. Silicon nanowires for Li-based battery anodes: A review. *J. Mater. Chem. A* 1: 9566–86.

35. Chan, C.K., Peng, H., Liu, G., McIlwrath, K., Zhang, X.F., Huggins, R.A., Cui, Y. 2008. High-performance lithium battery anodes using silicon nanowires. *Nat. Nanotechnol.* 3: 31–5.

36. Wu, H., Chan, G., Choi, J.W., Ryu, I., Yao, Y., McDowell, M.T., Lee, S.W., Jackson, A., Yang, Y., Hu, L., Cui, Y. 2012. Stable cycling of double-walled silicon nanotube battery anodes through solid electrolyte interphase control. *Nat. Nanotechnol.* 7: 310–15.

37. Deshpande, R., Cheng, Y.-T., Verbrugge, M.W. 2010. Modeling diffusion-induced stress in nanowire electrode structures. *J. Power Sources* 195: 5081–8.

38. Kim, H., Cho, J. 2008. Superior lithium electroactive mesoporous si@carbon core-shell nanowires for lithium battery anode material. *Nano Lett.* 8: 3688–91.

39. Liu, X.H., Zhong, L., Huang, S., Mao, S.X., Zhu, T., Huang, J.Y. 2012. Size-dependent fracture of silicon nanoparticles during lithiation. *ACS Nano* 6: 1522–31.

40. Hwa, Y., Kim, W.-S., Hong, S.-H., Sohn, H.-J. 2012. High capacity and rate capability

of core–shell structured nano-Si/C anode for Li-ion batteries. *Electrochim. Acta* 71: 201–05.

41. Tao, H.-C., Yang, X.-L., Zhang, L.-L., Ni, S.-B. 2014. Double-walled core-shell structured Si@SiO$_2$@C nanocomposite as anode for lithium-ion batteries. *Ionics* 20: 1547–52.

42. Chen, D., Mei, X., Ji, G., Lu, M., Xie, J., Lu, J., Lee, J.Y. 2012. Reversible lithium-ion storage in silver-treated nanoscale hollow porous silicon particles. *Angew. Chem. Int. Ed.* 51: 2409–13.

43. Murphy, D.W., Cava, R.J., Zahurak, S., Santoro, A. 1983. Ternary Li$_x$TiO$_2$ phases from insertion reactions. *Solid State Ionics* 9-10: 413–7.

44. Colbow, K.M., Dahn, J.R., Hearing, R.R. 1989. Structure and electrochemistry of spinel oxides LiTi$_2$O$_4$ and Li$_{4/3}$Ti$_{5/3}$O$_4$. *J. Power Sources* 26: 397–402.

45. Koshiba Nobuharu, Takata Kenichi, Asaka Emi, Nakanishi, Makoto. Rechargeable lithium cell and process for making an anode for use in the cell. August 13, 1996. US Patent No. 5545468.

46. Ohzuku, T., Ueda, A., Yamamoto, N. 1955. Zero-strain insertion material of Li[Li$_{1/3}$Ti$_{5/3}$]O$_4$ for rechargeable lithium cells. *J. Electrochem. Soc.* 142: 1431–5.

47. Ferg, E., Gummow, R.J., de Kock, A., Thackeray, M.M. 1994. Spinel anodes for lithium-ion batteries. *J. Electrochem. Soc.* 141: L147–L150.

48. Zaghib, K., Armand, M., Gauthier, M. 1998. Electrochemistry of anodes in solid-state Li-ion polymer batteries. *J. Electrochem. Soc.* 145: 3135–40.

49. Robertson, D., Trevino, L., Tukamoto, H., Irvine, J.T.S. 1999. New inorganic spinel oxides for use as negative electrode materials in future lithium-ion batteries. *J. Power Sources* 81-82: 352–7.

50. Ge, H., Li, N., Li, D., Dai, C., Wang, D. 2008. Study on the effect of Li doping in spinel Li$_{4+x}$Ti$_{5-x}$O$_{12}$ (0 < x < 0.2) materials for lithium-ion batteries. *Electrochem. Comm.* 10: 1031–4.

51. Yoshikawa, D., Kadoma, Y., Kim, J.-M., Ui, K., Kumagai, N., Kitamura, N., Idemoto, Y. 2010. Spray-drying synthesized lithium-excess Li$_{4+x}$Ti$_{5-x}$O$_{12-δ}$ and its electrochemical property as negative electrode material for Li-ion batteries. *Electrochim. Acta* 55: 1872–9.

52. Zhang, Q., Zhang, C., Li, B., Kang, S., Li, X., Wang, Y. 2013. Preparation and electrochemical properties of Ca-doped Li$_4$Ti$_5$O$_{12}$ as anode materials in lithium-ion battery. *Electrochim. Acta* 98: 146–152.

53. Wang, W., Jiang, B., Xiong, W., Wang, Z., Jiao, S. 2013. A nanoparticle Mg-doped Li$_4$Ti$_5$O$_{12}$ for high rate lithium-ion batteries. *Electrochim. Acta* 114: 198–204.

54. Kubiak, P., Garcia, A., Womes, M., Aldon, L., Olivier-Fourcade, J., Lippens, P.-E., Jumas, J.-C. 2003. Phase transition in the spinel Li$_4$Ti$_5$O$_{12}$ induced by lithium insertion. Influence of the substitutions Ti/V, Ti/Mn, Ti/Fe. *J. Power Sources* 119-121: 626–30.

55. Yi, T.-F., Shu, J., Zhu, Y.-R., Zhu, X.-D., Zhu, R.-S., Zhou, A.-N. 2010. Advanced electrochemical performance of Li$_4$Ti$_{4.95}$V$_{0.05}$O$_{12}$ as a reversible anode material down to 0V. *J. Power Sources* 195: 285–8.

56. Yu, Z., Zhang, X., Yang, G., Liu, J., Wang, J., Wang, R., Zhang, J. 2011. High rate capability and long-term cyclability of Li$_4$Ti$_{4.9}$V$_{0.1}$O$_{12}$ as anode material in lithium ion battery. *Electrochim. Acta* 56: 8611–8617.

57. Zhang, B., Du, H., Li, B., Kang, F. 2010. Structure and electrochemical properties of Zn-doped Li$_4$Ti$_5$O$_{12}$ as anode materials in Li-Ion battery. *Electrochem. Solid-State Lett.* 13: A36–A38.

58. Yi, T.-F., Liu, H., Zhu, Y.-R., Jiang, L.-J., Xie, Y., Zhu, R.-S. 2012. Improving the high rate performance of Li$_4$Ti$_5$O$_{12}$ through divalent zinc substitution. *J. Power Sources* 215: 258–65.

59. Xu, Y., Hong, Z., Xia, L., Yang, J., Wei, M. 2013. One step sol–gel synthesis of $Li_2ZnTi_3O_8$/C nanocomposite with enhanced lithium-ion storage properties. *Electrochim. Acta* 88: 74–78.

60. Hong, Z.S., Wei, M.D., Ding, X.K., Jiang, L.L., Wei, K.M. 2010. $Li_2ZnTi_3O_8$ nanorods: A new anode material for lithium-ion battery. *Electrochem. Comm.* 12: 720–3.

61. Wang, L., Wu, L.J., Li, Z.H., Lei, G.T., Xiao, Q.J., Zhang, P. 2011. Synthesis and electrochemical properties of $Li_2ZnTi_3O_8$ fibers as an anode material for lithium-ion batteries. *Electrochim. Acta* 56: 5343–6.

62. Hong, Z.S., Zheng, X.Z., Ding, X.K., Jiang, L.L., Wei, M.D., Wei, K.M. 2011. Complex spinel titanate nanowires for a high rate lithium-ion battery. *Energy Environ. Sci.* 4: 1886–91.

63. Zhang, Q., Zhang, C., Li, B., Jiang, D., Kang, S., Lia, X., Wang, Y. 2013. Preparation and characterization of W-doped $Li_4Ti_5O_{12}$ anode material for enhancing the high rate performance. *Electrochim. Acta* 107: 139–46.

64. Shahua, H., Zhaoyin, W., Zhonghua, G., Xiujian, Z. 2005 Preparation and cycling performance of Al^{3+} and F^- co-substituted compounds $Li_4Al_xTi_{5-x}F_yO_{12-y}$. *Electrochim. Acta* 50: 4057–62.

65. Wang, Z., Chen, G., Xu, J., Lv, Z., Yang, W. 2011. Synthesis and electrochemical performances of $Li_4Ti_{4.95}Al_{0.05}O_{12}$/C as anode material for lithium-ion batteries. *J. Phys. Chem. Solids* 72: 773–8.

66. Lin, J.-Y., Hsu, C.-C., Ho, H.-P., Wu, S-h. 2013. Sol–gel synthesis of aluminum doped lithium titanate anode material for lithium ion batteries. *Electrochim. Acta* 87: 126–32.

67. Shahua, H., Zhaoyin, W., Xiujian, Z., Zuxiang, L. 2005. Preparation and Electrochemical performance of spinel-type compounds $Li_4Al_yTi_{5-y}O_{12}$ (y = 0, 0.10, 0.15, 0.25). *J. Electrochem. Soc.* 152: A186–A190.

68. Cai, R., Jiang, S.M., Yu, X., Zhao, B.T., Wang, H.T., Shao, Z.P. 2012. A novel method to enhance rate performance of an Al-doped $Li_4Ti_5O_{12}$ electrode by post-synthesis treatment in liquid formaldehyde at room temperature. *J. Mater. Chem.* 22: 8013–21.

69. Park, J.S., Baek, S.-H., Jeong, Y.-I., Noh, B.-Y., Kim, J.H. 2013. Effects of a dopant on the electrochemical properties of $Li_4Ti_5O_{12}$ as a lithium-ion battery anode material. *J. Power Sources* 244: 527–31.

70. Stenina, I.A., Sobolev, A.N., Yaroslavtsev, S.A., Rusakov, V.S., Kulova, T.L., Skundin, A.M., Yaroslavtsev, A.B. 2016. Influence of iron doping on structure and electrochemical properties of $Li_4Ti_5O_{12}$. *Electrochimica Acta* 219: 524–30.

71. Kulova, T., Kuz'mina, A., Skundin, A., Stenina, I., Yaroslavtsev, A. 2017. Electrochemical behavior of gallium-doped lithium titanate in a wide range of potentials. *Int. J. Electrochem. Sci.* 12: 3197–211.

72. Lin, C., Lai, M.O., Lu, L., Zhou, H., Xin, Y. 2013. Structure and high rate performance of Ni^{2+} doped $Li_4Ti_5O_{12}$ for lithium ion battery. *J. Power Sources* 244: 272–9.

73. Xu, W., Chen, X., Wang, W., Choi, D., Ding, F., Zheng, J., Nie, Z., Choi, Y.J., Zhang, J.-G., Yang, Z.G. 2013. Simply AlF_3-treated $Li_4Ti_5O_{12}$ composite anode materials for stable and ultrahigh power lithium-ion batteries. *J. Power Sources* 236: 169–74.

74. Zhao, Z., Xu, Y., Ji, M., Zhang, H. 2013. Synthesis and electrochemical performance of F-doped $Li_4Ti_5O_{12}$ for lithium-ion batteries. *Electrochim. Acta* 109: 645–50.

75. Stenina, I.A., Kulova, T.L., Skundin, A.M., Yaroslavtsev, A.B. 2018. Effects of carbon coating from sucrose and PVDF on electrochemical performance of $Li_4Ti_5O_{12}$/C composites in different potential ranges. *J. Solid State Electrochem.* 22: 2631–9.

76. Junjie, H., Zhiyu, J. 2008. The preparation and characterization of $Li_4Ti_5O_{12}$/carbon nanotubes for lithium ion battery. *Electrochim. Acta* 53: 7756–9.

77. Li, X., Qu, M., Huai, Y., Yu, Z. 2010. Preparation and electrochemical performance

of $Li_4Ti_5O_{12}$/carbon/carbon nano-tubes for lithium ion battery. *Electrochim. Acta* 55: 2978–82.

78. Jhan, Y.R., Duh, J.G. 2012. Synthesis of entanglement structure in nanosized $Li_4Ti_5O_{12}$/ multi-walled carbon nanotubes composite anode material for Li-ion batteries by ball-milling-assisted solid-state reaction. *J. Power Sources* 198: 294–7.

79. Ni, H., Fan, L.-Z. 2012. Nano-$Li_4Ti_5O_{12}$ anchored on carbon nanotubes by liquid phase deposition as anode material for high rate lithium-ion batteries. *J. Power Sources* 214: 195–9.

80. Fang, W., Zuo, P., Ma, Y., Cheng, X., Liao, L., Yin, G. 2013. Facile preparation of $Li_4Ti_5O_{12}$/AB/MWCNTs composite with high-rate performance for lithium ion battery. *Electrochim. Acta* 94: 294–9.

81. Zhang, B., Liu, Y., Huang, Z., Oh, S., Yu, Y., Mai, Y.-W., Kim, J.-K. 2012. Urchin-like $Li_4Ti_5O_{12}$-carbon nanofiber composites for high rate performance anodes in Li-ion batteries. *J. Mater. Chem.* 22: 12133–40.

82. Shen, L., Yuan, C., Luo, H., Zhang, X., Xu, K., Zhang, F. 2011. In situ growth of $Li_4Ti_5O_{12}$ on multi-walled carbon nanotubes: Novel coaxial nanocables for high rate lithium ion batteries. *J. Mater. Chem.* 21: 761–7.

83. Zhu, N., Liu, W., Xue, M., Xie, Z., Zhao, D., Zhang, M., Chen, J., Cao, T. 2010. Graphene as a conductive additive to enhance the high-rate capabilities of electrospun $Li_4Ti_5O_{12}$ for lithium-ion batteries. *Electrochim. Acta* 55: 5813–8.

84. Shi, Y., Wen, L., Li, F., Cheng, H.-M. 2011. Nanosized $Li_4Ti_5O_{12}$/graphene hybrid materials with low polarization for high rate lithium ion batteries. *J. Power Sources* 196: 8610–7.

85. Ri, S.G., Zhan, L., Wang, Y., Zhou, L., Hu, J., Liu, H. 2013. $Li_4Ti_5O_{12}$/graphene nanostructure for lithium storage with high-rate performance. *Electrochim. Acta* 109389–94.

86. Stenina, I.A., Sobolev, A.N., Kuz'mina, A.A., Kulova, T.L., Skundin, A.M., Tabachkova, N.Yu., Yaroslavtsev, A.B. 2017. Electrochemical properties of $Li_4Ti_5O_{12}$/C and $Li_4Ti_5O_{12}$/C/Ag nanomaterials. *Inorg. Mater.* 53: 1039–45.

87. Stenina, I.A., Kulova, T.L., Skundin, A.M., Yaroslavtsev, A.B. 2016. High grain boundary density $Li_4Ti_5O_{12}$/anatase–TiO_2 nanocomposites as anode material for Li-ion batteries. *Mater. Res. Bull.* 75: 178–84.

88. Bresser, D., Paillard, E., Copley, M., Bishop, P., Winter, M., Passerini, S. 2012. The importance of "going nano" for high power battery materials. *J. Power Sources* 219: 217–22.

89. Lee, S.C., Lee, S.M., Lee, J.W., Lee, J.B., Lee, S.M., Han, S.S., Lee, H.C., Kim, H.J. 2009. Spinel $Li_4Ti_5O_{12}$ nanotubes for energy storage materials. *J. Phys. Chem. C* 113: 18420–3.

90. Jo, M.R., Jung, Y.S., Kang, Y.-M. 2012. Tailored $Li_4Ti_5O_{12}$ nanofibers without standing kinetics for lithium rechargeable batteries. *Nanoscale* 4: 6870–5.

91. Kim, J.-H., Khanal, S., Islam, M., Khatri, A., Choi, D. 2008. Electrochemical characterization of vertical arrays of tin nanowires grown on silicon substrates as anode materials for lithium rechargeable microbatteries. *Electrochem. Commun.* 10: 1688–90.

92. Yang, S., Zavalij, P.Y., Whittingham, M.S. 2003. Anodes for lithium batteries: Tin revisited. *Electrochem. Commun.* 5: 587–90.

93. Hu, R., Zeng, M., Ying, C., Li, V., Zhu, M.J. 2009. Microstructure and electrochemical performance of thin film anodes for lithium ion batteries in immiscible Al–Sn system. *J. Power Sources* 188: 268–73.

94. Hu, R.Z., Zeng, M.Q., Zhu, M. 2009. Cyclic durable high-capacity Sn/Cu_6Sn_5 composite thin film anodes for lithium ion batteries prepared by electron-beam evaporation deposition. *Electrochim. Acta* 54: 2843–50.

95. Mukaibo, H., Sumi, T., Yokoshima, T., Momma, T., Osaka, T. 2003. Electrodeposited Sn-Ni alloy film as a high capacity anode material for lithium-ion secondary batteries. *Electrochem. Solid-State Lett.* 6: A218–A220.

96. Hao, J., Wang, Y., Guo, Q., Zhao, J., Li, Y. 2019. Structural strategies for germanium-based anode materials to enhance lithium storage. *Part. Part. Syst. Charact.* 36(9): Article No. 1900248.

97. Chockla, A.M., Klavetter, K.C., Mullins, C.B., Korgel, B.A. 2012. Solution-grown germanium nanowire anodes for lithium-ion batteries. *ACS Appl. Mater. Interfaces* 4: 4658–64.

98. Meshgi, M.A., Biswas, S., McNulty, D., O'Dwyer, C., Verni, G.A., O'Connell, J., Davitt, F., Letofsky-Papst, I., Poelt, P., Holmes, J.D., Marschner, C. 2017. Rapid, low-temperature synthesis of germanium nanowires from oligosilyl germane precursors. *Chem. Mater.* 29: 4351–60.

99. Carim, A.I., Collins, S.M., Foley, J.M., Maldonado, S. 2011. Benchtop electrochemical liquid–liquid–solid growth of nanostructured crystalline germanium. *J. Am. Chem. Soc.* 133: 13292–5.

100. Mahenderkar, N.K., Liu, Y.-C., Koza, J.A., Switzer, J.A. 2014. Electrodeposited germanium nanowires. *ACS Nano* 8: 9524-30.

101. Ma, L., Fahrenkrug, E., Gerber, E., Crowe, A.J., Venable, F., Bartlett, B.M., Maldonado, S. 2017. High-performance polycrystalline Ge microwire film anodes for Li ion batteries, *ACS Energy Lett.* 2: 238–43.

102. Gavrilin, I.M., Smolyaninov, V.A., Dronov, A.A., Gavrilov, S.A., Trifonov, A.Yu., Kulova, T.L., Kuz'mina, A.A., Skundin, A.M. 2018. Study of the process of reversible insertion of lithium into nanostructured materials based on germanium. *Russ. J. Electrochem.* 54: 907–12.

103. Vassiliev, S.Yu., Yusipovich, A.I., Rogynskaya, Yu.E., Chibirova, F.Kh., Skundin, A.M., Kulova, T.L. 2005. Nanostructured SnO_2-TiO_2 films as related to lithium intercalation. *J. Solid State Electrochem.* 9: 698–705.

104. Mizushima, K., Jones, P.C., Wiseman, P.J., Goodenough, J.B. 1980. Li_xCoO_2 ($0 < x < 1$): A new cathode material for batteries of high energy density. *Mat. Res. Bull.* 15: 783–9.

105. Matsushita, T., Dokko, K., Kanamura, K. 2005. Comparison of electrochemical behavior of $LiCoO_2$ thin films prepared by sol–gel and sputtering processes. *J. Electrochem. Soc.* 152: A2229–A2237.

106. Rho, Y.H., Kanamura, K. 2004. Li^+-ion diffusion in $LiCoO_2$ thin film prepared by the poly(vinylpyrrolidone) sol–gel method. *J. Electrochem. Soc.* 151: A1406–A1411.

107. Bouwman, P.J., Boukamp, B.A., Bouwmeester, H.J.M., Wondergem, H.J., Notten, P.H.L. 2001. Structural analysis of submicrometer $LiCoO_2$ films. *J. Electrochem. Soc.* 148: A311–A317.

108. Paulsen, J.M., Mueller-Neuhaus, J.R., Dahn, J.R. 2000. Layered $LiCoO_2$ with a different oxygen stacking (O_2 structure) as a cathode material for rechargeable lithium batteries. *J. Electrochem. Soc.* 147: 508–16.

109. Bates, J.B., Dudney, N.J., Neudecker, B.J., Hart, F.X., Jun, H.P., Hackney, S.A. 2000. Preferred orientation of polycrystalline $LiCoO_2$ films. *J. Electrochem. Soc.* 147: 59–70.

110. Kanamura, K., Goto, A., Ho, R.Y., Umegaki, T., Toyoshima, K., Okada, K.-i., Hakuta, Y., Adschiri, T., Arai, K. 2000. Preparation and electrochemical characterization of $LiCoO_2$ particles prepared by supercritical water synthesis. *Electrochem. Solid-State Lett.* 3: 256–8.

111. Endo, E., Yasuda, T., Kita, A., Yamaura, K., Sekai, K. 2000. A $LiCoO_2$ cathode modified by plasma chemical vapor deposition for higher voltage performance. *J. Electrochem. Soc.* 147: 1291–4.

112. Cho, J., Kim, C.-S., Yoo, S.-I. 2000. Improvement of structural stability of $LiCoO_2$ cathode during electrochemical cycling by sol–gel coating of SnO_2. *Electrochem. Solid-State Lett.* 3: 362–5.

113. Callister, W.D. Jr. 1997. *Materials Science and Engineering: An Introduction.* 4th ed., Appendix C. p. 787. John Wiley & Sons: New York.

114. Fredel, M.C., Boccaccini, A.R. 1996. Processing and mechanical properties of biocompatible Al_2O_3 platelet-reinforced TiO_2. *J. Mater. Sci.* 31: 4375–80.

115. Sehgal, J., Ito, S. 1999. Brittleness of glass. *J. Non-Cryst. Solids* 253: 126–32.

116. Cho, J., Kim, G. 1999. Enhancement of thermal stability of $LiCoO_2$ by $LiMn_2O_4$ coating. *Electrochem. Solid-State Lett.* 2: 253–5.

117. Cho, J., Kim, Y.I., Park, B. 2000. Novel $LiCoO_2$ cathode material with Al_2O_3 coating for a Li ion cell. *Chem. Mater.* 12: 3788–91.

118. Cho, J., Kim, Y.I., Park, B. 2001. $LiCoO_2$ cathode material that does not show a phase transition from hexagonal to monoclinic phase. *J. Electrochem. Soc.* 148: A1110–A1115.

119. Cho, J., Kim, Y.I., Kim, T.J., Park, B. 2001. Zero-strain intercalation cathode for rechargeable Li-ion cell. *Angew. Chem. Int. Ed. Engl.* 40: 3367–9.

120. Liu, Z., Wang, H., Fang, L., Lee, J.Y., Gan, L.M. 2002. Improving the high-temperature performance of $LiMn_2O_4$ spinel by micro-emulsion coating $LiCoO_2$. *J. Power Sources* 104: 101–7.

121. Kannan, A.M., Rabenberg, L., Manthiram, A. 2003. High capacity surface-modified $LiCoO_2$ cathodes for lithium-ion batteries. *Electrochem. Solid-State Lett.* 6: A16–A18.

122. Zhu, J.P., Xu, Q.B., Yang, H.W., Zhao, J.J., Yang, G. 2011. Recent development of $LiNi_{1/3}Co_{1/3}Mn_{1/3}O_2$ as cathode material of lithium ion battery. *J. Nanosci. Nanotechnol.* 11: 10357–68.

123. Madhavi, S., Subba Rao, G.V., Chowdari, B.V.R., Li, S.F.Y. 2001. Effect of aluminium doping on cathodic behaviour of $LiNi_{0.7}Co_{0.3}O_2$. *J. Power Sources* 93: 156–62.

124. Chen, C.H., Liu, J., Stoll, M.E., Henriksen, G., Vissers, D.R., Amine, K. 2004. Aluminum-doped lithium nickel cobalt oxide electrodes for high-power lithium-ion batteries. *J. Power Sources* 128: 278–85.

125. Kim, J., Lee, H., Cha, H., Yoon, M., Park, M., Cho, J. 2018. Prospect and reality of Ni-rich cathode for commercialization. *Adv. Energy Mater.* 8: Article No. 1702028.

126. Lai, Y.Q., Xu, M., Zhang, Z.A., Gao, C.H., Wang, P., Yu, Z.Y. 2016. Optimized structure stability and electrochemical performance of $LiNi_{0.8}Co_{0.15}Al_{0.05}O_2$ by sputtering nanoscale ZnO film. *J. Power Sources* 309: 20–6.

127. Cho, Y., Lee, Y.S., Park, S.A., Lee, Y., Cho, J. 2010. $LiNi_{0.8}Co_{0.15}Al_{0.05}O_2$ cathode materials prepared by TiO_2 nanoparticle coatings on $Ni_{0.8}Co_{0.15}Al_{0.05}(OH)_2$ precursors, *Electrochim. Acta* 56: 333–9.

128. Xu, Y., Li, X.H., Wang, Z.X., Guo, H.J., Peng, W.J., Pan, W. 2016. The enhanced high cutoff voltage electrochemical performances of $LiNi_{0.5}Co_{0.2}Mn_{0.3}O_2$ by the CeO_2 modification. *Electrochim. Acta* 219: 49–60.

129. Kim, H.-B., Park, B.-C., Myung, S.-T., Amine, K., Prakash, J., Sun, Y.-K. 2008. Electrochemical and thermal characterization of AlF_3-coated $Li[Ni_{0.8}Co_{0.15}Al_{0.05}]O_2$ cathode in lithium-ion cells. *J. Power Sources* 179: 347–50.

130. Liu, W.M., Tang, X., Qin, M.L., Li, G.L., Deng, J.Y., Huang, X.W. 2016. FeF_3-coated $LiNi_{0.8}Co_{0.15}Al_{0.05}O_2$ cathode materials with improved electrochemical properties. *Mater. Lett.* 185: 96–9.

131. Chen, C., Xu, M., Zhang, K., An, H., Zhang, G., Hong, B., Li, J., Lai, Y. 2019. Atomically ordered and epitaxially grown surface structure in core-shell $NCA/NiAl_2O_4$ enabling high voltage cyclic stability for cathode application. *Electrochim. Acta* 300: 437–44.

132. Li, Y., Yu, H., Hu, Y., Jiang, H., Li, C. 2018. Surface-engineering of layered $LiNi_{0.815}Co_{0.15}Al_{0.035}O_2$ cathode material for high-energy and stable Li-ion batteries. *J. Energy Chem.* 27: 559–64.

133. Sun, Y.K., Chen, Z., Noh, H.J., Lee, D.J., Jung, H.G., Ren, Y., Wang, S., Yoon, C.S., Myung, S.T., Amine, K. 2012. Nanostructured high-energy cathode materials for advanced lithium batteries. *Nat. Mater.* 11: 942–7.

134. Jo, M., Noh, M., Oh, P., Kim, Y., Cho, J. 2014. A new high power $LiNi_{0.81}Co_{0.1}Al_{0.09}O_2$ cathode material for lithium-ion batteries. *Adv. Energy Mater.* 4: Article No. 1301583.

135. Baghbanzadeh, M., Carbone, L., Cozzoli, P.D., Kappe, C.O. 2011. Microwave-assisted synthesis of colloidal inorganic nanocrystals. *Angew. Chem.* 50: 11312–59.

136. Ren, H., Wang, Y., Li, D., Ren, L., Peng, Z., Zhou, Y. 2008. Synthesis of $LiNi_{1/3}Co_{1/3}Mn_{1/3}O_2$ as a cathode material for lithium battery by the rheological phase method. *J. Power Sources* 178: 439–44.

137. Reddy, M.V., Subba Rao, G.V., Chowdari, B.V.R. 2006. Synthesis by molten salt and cathodic properties of $Li(Ni_{1/3}Co_{1/3}Mn_{1/3})O_2$. *J. Power Sources* 159: 263–7.

138. Padhi, A.K., Nanjundaswamy, K.S., Masquelier, C., Okada, S., Goodenough, J.B. 1997. Effect of structure on the Fe^{3+}/Fe^{2+} redox couple in iron phosphates. *J. Electrochem. Soc.* 144: 1609–13.

139. Franger, S., Le Cras, F., Bourbon, C., Rouault, H. 2002. $LiFePO_4$ synthesis routes for enhanced electrochemical performance. *Electrochem. Solid-State Lett.* 5: A231–A233.

140. Huang, H., Yin, S.-C., Nazar, L.F. 2001. Approaching theoretical capacity of $LiFePO_4$ at room temperature at high rates. *Electrochem. Solid-State Lett.* 4: A170–A172.

141. Dong, Y.Z., Zhao, Y.M., Duan, H., Chen, L., He, Z.F., Chen, Y.H. 2010. Electrochemical properties of single-phase $LiFePO_4$ synthesized using LiF as Li precursor and hydrogen and carbon gel as reducing agents. *J. Solid State Electrochem.* 14: 131–7.

142. Dong, Y.Z., Zhao, Y.M., Chen, Y.H., He, Z.F., Kuang, Q. 2009. Optimized carbon-coated $LiFePO_4$ cathode material for lithium-ion batteries. *Mater. Chem. Phys.* 115: 245–50.

143. Kim, J.-K., Cheruvally, G., Ahn, J.-H. 2008. Electrochemical properties of $LiFePO_4/C$ synthesized by mechanical activation using sucrose as carbon source. *J. Solid State Electrochem.* 12: 799–805.

144. Kwon, S.J., Kim, C.W., Jeong, W.T., Sub, K. 2004. Synthesis and electrochemical properties of olivine $LiFePO_4$ as a cathode material prepared by mechanical alloying. *J. Power Sources* 137: 93–9.

145. Kim, J.-K., Cheruvally, G., Choi, J.W., Kim, J.U., Ahn, J.H., Cho, G.B., Kim, K.W., Ahn, H.J. 2007. Effect of mechanical activation process parameters on the properties of $LiFePO_4$ cathode material. *J. Power Sources* 166: 211–8.

146. Kim, J.-K., Choi, J.W., Cheruvally, G., Kim, J.U., Ahn, J.H., Cho, G.B., Kim, K.W., Ahn, H.J. 2007. A modified mechanical activation synthesis for carbon-coated $LiFePO_4$ cathode in lithium batteries. *Mater. Lett.* 61: 3822–5.

147. Doeff, M.M., Wilcox, J.D., Yu, R., Aumentado, A., Marcinek, M., Kostecki, R. 2008. Impact of carbon structure and morphology on the electrochemical performance of $LiFePO_4/C$ composites. *J. Solid State Electrochem.* 12: 995–1001.

148. Hu, Y., Doeff, M.M., Kostecki, R., Finones, R. 2004. Electrochemical performance of sol-gel synthesized $LiFePO_4$ in lithium batteries. *J. Electrochem. Soc.* 151: A1279–A1285.

149. Aurbach, D., Markovsky, B., Salitra, G., Markevich, E., Talyossef, Y., Koltypin, M., Nazar, L., Ellis, B., Kovacheva, D. 2007. Review on electrode–electrolyte solution interactions, related to cathode materials for Li-ion batteries. *J. Power Sources* 165: 491–9.

150. Zhang, J., Gao, L. 2004. Synthesis of antimony-doped tin oxide (ATO) nanoparticles by the nitrate–citrate combustion method. *Mater. Res. Bull.* 39: 2249–55.

151. Chick, L.A., Pederson, L.R., Maupin, G.D., Bates, J.L., Thomas, L.E., Exarhos, G.J. 1990. Glycine-nitrate combustion synthesis of oxide ceramic powders. *Mater. Lett.* 10: 6–12.

152. Junior, A.F., de Olivera Lima, E.C., Novak, M.A., Wells, P.R. 2007. Synthesis of nanoparticles of $Co_xFe_{(3-x)}O_4$ by combustion reaction method. *J. Magnetism Magnetic Mater.* 308: 198–202.

153. Chen, S-Y., Gao, B., Su, L-H., Mi, C-H., Zhang, X-G. 2009. Electrochemical properties of $LiFePO_4$/C synthesized using polypyrrole as carbon source. *J. Solid State Electrochem.* 13: 1361–6.

154. Fey, G.T.-K., Lu, T.-L. 2008. Morphological characterization of $LiFePO_4$/C composite cathode materials synthesized via a carboxylic acid route. *J. Power Sources* 178: 807–14.

155. Li, X., Wang, W., Shi, Ch., Wang, H., Xing, Y. 2009. Structural and electrochemical characterization of $LiFePO_4$/C prepared by a sol–gel route with long- and short-chain carbon sources. *J. Solid State Electrochem.* 13: 921–6.

156. Wu, L., Li, X., Wang, Z., Li, L., Zheng, J., Guo, H., Hu, Q., Fang, J. 2009. Synthesis and electrochemical properties of metals-doped $LiFePO_4$ prepared from the $FeSO_4 \cdot 7H_2O$ waste slag. *J. Power Sources* 189: 681–4.

157. Wu, Sh., Chen, M.-S., Chien, Ch.-J., Fu, T.-P. 2009. Preparation and characterization of Ti^{4+}-doped $LiFePO_4$ cathode materials for lithium-ion batteries. *J. Power Sources* 189: 440–4.

158. Sun, C.S., Zhou, Z., Xu, Z.G., Wang, D.G., Wei, J.P., Bian, X.K., Yan, J. 2009. Improved high-rate charge/discharge performances of $LiFePO_4$/C via V-doping. *J. Power Sources* 193: 841–5.

159. Liu, Z., Zhang, X., Hong, L. 2009. Preparation and electrochemical properties of spherical $LiFePO_4$ and $LiFe_{0.9}Mg_{0.1}PO_4$ cathode materials for lithium rechargeable batteries. *J. Appl. Electrochem.* 39: 2433–8.

160. Lee, S.B., Cho, S.H., Heo, J.B., Aravindan, V., Kim, H.S., Lee, Y.S. 2009. Copper-substituted, lithium rich iron phosphate as cathode material for lithium secondary batteries. *J. Alloys Comp.* 488: 380–5.

161. Shenouda, A.Y., Liu, H.K. 2009. Studies on electrochemical behaviour of zinc-doped $LiFePO_4$ for lithium battery positive electrode. *J. Alloys Comp.* 477: 498–503.

162. Wang, Y., Yang, Y., Hu, X., Yang, Y., Shao, H. 2009. Electrochemical performance of Ru-doped $LiFePO_4$/C cathode material for lithium-ion batteries. *J. Alloys Comp.* 481: 590–4.

163. Nam, K-W., Yoon, W-S., Zaghib, K., Chung, K.Y., Yang, X-Q. 2009. The phase transition behaviors of $Li_{1-x}Mn_{0.5}Fe_{0.5}PO_4$ during lithium extraction studied by in situ X-ray absorption and diffraction techniques. *Electrochem. Commun.* 11: 2023–6.

164. Liu, H., Xie, J. 2009. Synthesis and characterization of $LiFe_{0.9}Mg_{0.1}PO_4$/nano-carbon webs composite cathode. *J. Mater. Process. Technol.* 209: 477–81.

165. Li, L., Li, X., Wang, Z., Wu, L., Zheng, J., Guo, H. 2009. Stable cycle-life properties of Ti-doped $LiFePO_4$ compounds synthesized by co-precipitation and normal temperature reduction method. *J. Phys. Chem. Solids* 70: 238–42.

166. Bini, M., Mozzati, M.C., Galinetto, P., Capsoni, D., Ferrari, S., Grandi, M.S., Massarotti, V. 2009. Structural, spectroscopic and magnetic investigation of the $LiFe_{1-x}Mn_xPO_4$ (x = 0–0.18) solid solution. *J. Solid State Chem.* 182: 1972–81.

167. Arumugam, D., Paruthimal Kalaignan, G., Manisankar, P. 2009. Synthesis and electrochemical characterizations of nano-crystalline $LiFePO_4$ and Mg-doped $LiFePO_4$

cathode materials for rechargeable lithium-ion batteries. *J. Solid State Electrochem.* 13: 301–7.

168. Guo, Z.P., Liu, H., Bewlay, S., Liu, H.K., Dou, S.X. 2005. Start-fine-particle carbon-enriched $Li_{0.98}Mg_{0.02}FePO_4$ synthesized by a novel modified solid-state reaction. *Synthetic Metals* 153: 113–6.

169. Hong, J., Wang, C., Kasavajjula, U. 2006. Kinetic behavior of $LiFeMgPO_4$ cathode material for Li-ion batteries. *J. Power Sources* 162: 1289–96.

170. Lee, K.T., Lee, K.S. 2009. Electrochemical properties of $LiFe_{0.9}Mn_{0.1}PO_4/Fe_2P$ cathode material by mechanical alloying. *J. Power Sources* 189: 435–9.

171. Islam, M.S., Driscoll, D.J., Fisher, C.A.J., Slater, P.R. 2005. Atomic-scale investigation of defects, dopants, and lithium transport in the $LiFePO_4$ olivine-type battery material. *Chem. Mater.* 17: 5085–92.

172. Baddour-Hadjean, R., Navone, C., Pereira-Ramos, J.P. 2009. In situ Raman microspectrometry investigation of electrochemical lithium intercalation into sputtered crystalline V_2O_5 thin films. *Electrochim. Acta* 54: 6674–9.

173. Navone, C., Baddour-Hadjean, R., Pereira-Ramos, J.P., Salot, R. 2008. A kinetic study of electrochemical lithium insertion into oriented V_2O_5 thin films prepared by RF sputtering. *Electrochim. Acta* 53: 3329–36.

174. Zhang, J.-G., McGraw, J.M., Turner, J., Ginley, D. 1997. Charging capacity and cycling stability of VO_x films prepared by pulsed laser deposition. *J. Electrochem. Soc.* 144: 1630–4.

175. Gallasch, T., Stockhoff, T., Baither, D., Schmitz, G. 2011. Ion beam sputter deposition of V_2O_5 thin films. *J. Power Sources* 196: 428–35.

176. Groult, H., Le Van, K., Mantoux, A., Perrigaud, L., Doppelt, P. 2007. Study of the Li^+ insertion into V_2O_5 films deposited by CVD onto various substrates. *J. Power Sources* 174: 312–20.

177. Le Van, K., Groult, H., Mantoux, A., Perrigaud, L., Lantelme, F., Lindström, R., Badour-Hadjean, R., Zanna, S., Lincot, D. 2006. Amorphous vanadium oxide films synthesised by ALCVD for lithium rechargeable batteries. *J. Power Sources* 160: 592–601.

178. Varadaraajan, V., Satishkumar, B.C., Nanda, J., Mohanty, P. 2011. Direct synthesis of nanostructured V_2O_5 films using solution plasma spray approach for lithium battery application. *J. Power Sources* 196: 10704–11.

179. Terukov, E.I., Nikitin, S.E., Nikolaev, Yu.A., Kulova, T.L., Skundin, A.M. 2009. Lithium incorporation into thin films of vanadium oxides. *Techn. Phys. Lett.* 35: 1111–3.

180. Park, H.-K., Smyrl, W.H., Ward, M.D. 1995. V_2O_5 Xerogel films as intercalation hosts for lithium: I. Insertion stoichiometry, site concentration, and specific energy. *J. Electrochem. Soc.* 142: 1068–73.

181. Vivier, V., Farcy, J., Pereira-Ramos, J.P. 1998. Electrochemical lithium insertion in sol-gel crystalline vanadium pentoxide thin films. *Electrochim. Acta* 44: 831–9.

182. Ban, C., Chernova, N.A., Whittingham, M.S. 2009. Electrospun nano-vanadium pentoxide cathode. *Electrochem. Comm.* 11: 522–5.

183. Li, X., Fu, J., Pan, Z., Su, J., Xu, J., Gao, B., Peng, X., Wang, L., Zhang, X., Chu, P.K. 2016. Peapod-like V_2O_3 nanorods encapsulated into carbon as binder-free and flexible electrodes in lithium-ion batteries. *J. Power Sources* 331: 58–66.

184. O'Dwyer, C., Lavayen, V., Newcomb, S.B., Benavente, E., Santa Ana, M.A., González, G., Sotomayor Torres, C.M. 2007. Atomic layer structure of vanadium oxide nanotubes grown on nanourchin structures. *Electrochem. Solid-State Lett.* 10: A111–A114.

185. Cui, C.-j., Wu, G.-m., Shen, J., Zhou, B., Zhang, Z.-h., Yang, H.-y., She, S.-f. 2011. Synthesis and electrochemical performance of lithium vanadium oxide nanotubes as cathodes for rechargeable lithium-ion batteries. *Electrochim. Acta* 55: 2536–41.

186. Seng, K.H., Liu, J., Guo, Z.P., Chen, Z.X., Jia, D., Liu, H.K. 2011. Free-standing V_2O_5 electrode for flexible lithium ion batteries. *Electrochem. Comm.* 13: 383–6.

187. Yu, J., Liu, S., Cheng, B., Xiong, J., Yu, Y., Wang, J. 2006. Polymer-directed large-scale synthesis of single-crystal vanadium oxide nanobelts. *Mater. Chem. Phys.* 95: 206–10.

188. Semenenko, D.A., Itkis, D.M., Pomerantseva, E.A., Goodilin, E.A., Kulova, T.L., Skundin, A.M., Tretyakov, Y.D. 2010. $Li_xV_2O_5$ nanobelts for high capacity lithium-ion battery cathodes. *Electrochem. Comm.* 12: 1154–7.

189. Semenenko, D.A., Kozmenkova, A.Y., Itkis, D.M., Goodilin, E.A., Kulova, T.L., Skundin, A.M., Tretyakov, Y.D. 2012. Growth of thin vanadia nanobelts with improved lithium storage capacity in hydrothermally aged vanadia gels. *Cryst Eng Comm.* 14: 1561–7.

190. Feng, Y., Hou, F., Li, Y.L. 2009. A new low-temperature synthesis and electrochemical properties of LiV_3O_8 hydrate as cathode material for lithium-ion batteries. *J. Power Sources* 192: 708–13.

191. Fergus, J.W. 2010. Ceramic and polymeric solid electrolytes for lithium-ion batteries. *J. Power Sources* 195: 4554–69.

192. Knauth, P. 2009. Inorganic solid Li ion conductors: An overview. *Solid State Ionics* 180: 911–6.

193. Thangadurai, V., Weppner, W. 2006. Recent progress in solid oxide and lithium ion conducting electrolytes research. *Ionics* 12: 81–92.

194. Li, Q., Chen, J., Fan, L., Kong, X., Lu, Y. 2016. Progress in electrolytes for rechargeable Li-based batteries and beyond. *Green Energy & Environment* 1: 18–42.

195. Quartarone, E., Mustarelli, P. 2011. Electrolytes for solid-state lithium rechargeable batteries: Recent advances and perspectives. *Chem. Soc. Rev.* 40: 2525–40.

196. Teng, S., Tan, J., Tiwari, A. 2014. Recent developments in garnet based solid state electrolytes for thin film batteries. *Curr. Opin. Solid State Mater. Sci.* 18: 29–38.

197. Thangadurai, V., Narayanan, S., Pinzaru, D. 2014. Garnet-type solid-state fast Li ion conductors for Li batteries: Critical review. *Chem. Soc. Rev.* 43: 4714–27.

198. Zeier, W.G. 2014. Structural limitations for optimizing garnet-type solid electrolytes: A perspective. *Dalton Trans.* 43: 16133–8.

199. Bachman, J.C., Muy, S., Grimaud, A., Chang, H.-H., Pour, N., Lux, S.F., Paschos, O., Maglia, F., Lupart, S., Lamp, P., Giordano, L., Shao-Horn, Y. 2016. Inorganic solid-state electrolytes for lithium batteries: Mechanisms and properties governing ion conduction. *Chem. Rev.* 116: 140–62.

200. Tatsumisago, M., Nagao, M., Hayashi, A. 2013. Recent development of sulfide solid electrolytes and interfacial modification for all-solid-state rechargeable lithium batteries. *J. Asian Ceram. Soc.* 1: 17–25.

201. Liu, D., Zhu, W., Feng, Z., Guerfi, A., Vijh, A., Zaghib, K. 2016. Recent progress in sulfide-based solid electrolytes for Li-ion batteries. *Mat. Sci. Eng. B* 213: 169–76.

202. Kamaya, N., Homma, K., Yamakawa, Y., Hirayama, M., Kanno, R., Yonemura, M., Kamiyama, T., Kato, Y., Hama, S., Kawamoto, K. 2011. A lithium superionic conductor. *Nat. Mater.* 10: 682–6.

203. Minami, T., Hayashi, A., Tatsumisago, M. 2006. Recent progress of glass and glass-ceramics as solid electrolytes for lithium secondary batteries. *Solid State Ionics* 177: 2715–20.

204. Kim, Y.G., Wadley, H.N.G. 2009. Plasma-assisted deposition of lithium phosphorus oxynitride films: Substrate bias effects. *J. Power Sources* 187: 591–8.

205. Kim, Y.G., Wadley, H.N.G. 2008. Lithium phosphorous oxynitride films synthesized by a plasma-assisted directed vapor deposition approach. *J. Vac. Sci. Technol. A* 26: 174–84.

206. Vereda, F., Clay, N., Gerouki, A., Goldner, R.B., Haas, T., Zerigian, P. 2000. A study of electronic shorting in IBDA-deposited Lipon films. *J. Power Sources* 89: 201–5.

207. Yu, J.X., Bates, B., Jellison G.E., Hart, F.X. 1997. A stable thin-film lithium electrolyte: Lithium phosphorus oxynitride. *J. Electrochem. Soc.* 144: 524–32.

208. Hu, Z., Li, D., Xie, K. 2008. Influence of radio frequency power on structure and ionic conductivity of LiPON thin films. *Bull. Mater. Sci.* 31: 681–6.

209. Park, H.Y., Nam, S.C., Lim, Y.C., Choi, K.G., Lee, K.C., Park, G.B., Lee, S.-R., Kim, H.P., Cho, S.B. 2006. Effects of sputtering pressure on the characteristics of lithium ion conductive lithium phosphorous oxynitride thin film. *J. Electroceram.* 17: 1023–30.

210. Hamon, Y., Douard, A., Sabary, F., Marcel, C., Vinatier, P., Pecquenard, B., Levasseur, A. 2006. Influence of sputtering conditions on ionic conductivity of LiPON thin films. *Solid State Ionics* 177: 257–61.

211. Joo, K.H., Sohn, H.J., Vinatier, P., Pecquenard, B., Levasseur, A. 2004. Lithium ion conducting lithium sulfur oxynitride thin film. *Electrochem. Solid State Lett.* 7: A256–A258.

212. Jones, S.D., Akridge, J.R. 1994. Thin film rechargeable Li batteries. *Solid State Ionics* 69: 357–68.

213. Tatsumisago, M. 2004. Glassy materials based on Li_2S for all-solid-state lithium secondary batteries. *Solid State Ionics* 175: 13–8.

214. Mercier, R., Malugani, J.-P., Fahys, B., Robert, G. 1981. Superionic conduction in $Li_2S–P_2S_5–LiI$ – glasses. *Solid State Ionics* 5: 663–6.

215. Mizuno, F., Hayashi, A., Tadanaga, K., Tatsumisago, M. 2005. New, highly ion-conductive crystals precipitated from $Li_2S–P_2S_5$ glasses. *Adv. Mater.* 17: 918–21.

216. Hayashi, A., Muramatsu, H., Ohtomo, T., Hama, S., Tatsumisago, M. 2014. Improved chemical stability and cyclability in $Li_2S–P_2S_5–P_2O_5–ZnO$ composite electrolytes for all-solid-state rechargeable lithium batteries. *J. Alloy. Compd.* 591: 247–50.

217. Ohtomo, T., Hayashi, A., Tatsumisago, M., Kawamoto, K. 2013. Characteristics of the $Li_2O–Li_2S–P_2S_5$ glasses synthesized by the two-step mechanical milling. *J. Non-Cryst. Solids* 364: 57–61.

218. Trevey, J.E., Gilsdorf, J.R., Miller, S.W., Lee, S.-H. 2012. $Li_2S–Li_2O–P_2S_5$ solid electrolyte for all-solid-state lithium batteries. *Solid State Ionics* 214: 25–30.

219. Ujiie, S., Hayashi, A., Tatsumisago, M. 2012. Structure, ionic conductivity and electrochemical stability of $Li_2S–P_2S_5–LiI$ glass and glass–ceramic electrolytes. *Solid State Ionics* 211: 42–5.

220. Ujiie, S., Hayashi, A., Tatsumisago, M. 2013. Preparation and ionic conductivity of $(100–x)(0.8Li_2S \cdot 0.2P_2S_5) \cdot xLiI$ glass–ceramic electrolytes. *J. Solid State Electrochem.* 17: 675–80.

221. Zhang, Z., Kennedy, J.H. 1990. Synthesis and characterization of the $B_2S_3–Li_2S$, the $P_2S_5–Li_2S$ and the $B_2S_3–P_2S_5–Li_2S$ glass systems. *Solid State Ionics* 38: 217–24.

222. Fu, J. 1997. Superionic conductivity of glass-ceramics in the system $Li_2O–Al_2O_3–TiO_2–P_2O_5$. *Solid State Ionics* 96: 195–200.

223. Fu, J. 1997. Fast Li ion conducting glass-ceramics in the system $Li_2O–Al_2O_3–GeO_2–P_2O_5$. *Solid State Ionics* 104: 191–4.

224. Mizuno, F., Hayashi, A., Tadanaga, K., Tatsumisago, M. 2006. High lithium ion conducting glass-ceramics in the system $Li_2S–P_2S_5$. *Solid State Ionics* 177: 2721–5.

225. Seino, Y., Ota, T., Takada, K., Hayashi, A., Tatsumisago, M. 2014. A sulphide lithium super ion conductor is superior to liquid ion conductors for use in rechargeable batteries. *Energy Environ. Sci.* 7: 627–31.

226. Stramare, S., Thangadurai, V., Weppner. W. 2003. Lithium lanthanum titanates: A review. *Chem. Mater.* 15: 3974–90.

227. Bohnke, O. 2008. The fast lithium-ion conducting oxides $Li_{3x}La_{2/3-x}TiO_3$ from fundamentals to application. *Solid State Ionics* 179: 9–15.

228. Alonso, J.A., Sanz, J., Santamaria, J., Leon, C., Varez, A., Fernandez-Diaz, M.T. 2000. On the location of Li^+ cations in the fast Li-cation conductor $La_{0.5}Li_{0.5}TiO_3$ Perovskite. *Angew. Chem. Int. Ed.* 39: 619–21.

229. Ahn, J.-K., Yoon, S.-G. 2005. Characteristics of amorphous lithium lanthanum titanate electrolyte thin films grown by PLD for use in rechargeable lithium microbatteries. *Electrochem. Solid-State Lett.* 8: A75–A78.

230. Kanno, R., Murayama, M. 2001. Lithium ionic conductor thio-LISICON: The Li_2S–GeS_2–P_2S_5 system. *J. Electrochem. Soc.* 148: A742–A746.

231. Takada, K., Inada, T., Kajiyama, A., Sasaki, H., Kondo, S., Watanabe, M., Murayama, M., Kanno, R. 2003. Solid-state lithium battery with graphite anode. *Solid State Ionics* 158: 269–74.

232. Inada, T., Kobayashi, T., Sonoyama, N., Yamada, A., Kondo, S., Nagao, M., Kanno, R. 2009. All solid-state sheet battery using lithium inorganic solid electrolyte, thio-LISICON. *J. Power Sources* 194: 1085–8.

233. Kamaya, N., Homma, K., Yamakawa, Y., Hirayama, M., Kanno, R., Yonemura, M., Kamiyama, T., Kato, Y., Hama, S., Kawamoto, K., Matsui, A. 2011. A lithium superionic conductor. *Nat. Mater.* 10: 682–6.

234. Murugan, R., Weppner, W., Schmid-Beurmann, P., Thangadurai, V. 2007. Structure and lithium ion conductivity of bismuth containing lithium garnets $Li_5La_3Bi_2O_{12}$ and $Li_6SrLa_2Bi_2O_{12}$. *Mater Sci Eng B* 143: 14–20.

235. Ohta, S., Kobayashi, T., Asaoka, T. 2011. High lithium ionic conductivity in the garnet type oxide $Li_{7-x}La_3(Zr_{2-x}, Nb_x)O_{12}$. *J. Power Sources* 196: 3342–5.

236. El Shinawi, H., Janek, J. 2013. Stabilization of cubic lithium-stuffed garnets of the type "$Li_7La_3Zr_2O_{12}$" by addition of gallium. *J. Power Sources* 225: 13–9.

237. Allen, J.L., Wolfenstine, J., Rangasamy, E., Sakamoto, J. 2012. Effect of substitution (Ta, Al, Ga) on the conductivity of $Li_7La_3Zr_2O_{12}$. *J. Power Sources* 206: 315–9.

238. Murugan, R., Thangadurai, V., Weppner, W. 2007. Fast lithium ion conduction in garnet-type $Li_7La_3Zr_2O_{12}$. *Angew. Chem. Int. Ed.* 46: 7778–81.

239. Shimonishi, Y., Toda, A., Zhang, T., Hirano, A., Imanishi, N., Yamamoto, O., Takeda, Y. 2011. Synthesis of garnet-type $Li_{7-x}La_3Zr_2O_{12-x}$ and its stability in aqueous solutions. *Solid State Ionics* 183: 48–53.

240. Kokal, I., Somer, M., Notten, P.H.L., Hintzen, H.T. 2011. Sol–gel synthesis and lithium ion conductivity of $Li_7La_3Zr_2O_{12}$ with garnet-related type structure. *Solid State Ionics* 185: 42–6.

241. Geiger, C.A., Alekseev, E., Lazic, B., Fisch, M., Armbruster, T., Langner, R., Fechtelkord, M., Kim, N., Pettke, T.T., Weppner, W. 2010. Crystal chemistry and stability of "$Li_7La_3Zr_2O_{12}$" garnet: A fast lithium-ion conductor. *Inorg Chem.* 50: 1089–97.

CHAPTER 3

PVD Methods for Manufacturing All-Solid-State Thin-Film Lithium-Ion Batteries

3.1. General aspects

3.1.1 Requirements for the functional layers of all-solid-state thin-film battery

Before considering the general PVD methods of film deposition, several features of the deposition of functional layers of all-solid-state thin-film lithium-ion batteries will be formulated. By features, one could assume the structure, elemental composition, conductivity and some other properties of the functional layers.

Analogs for all functional layers of all-solid-state lithium-ion batteries are their layers in batteries with a liquid electrolyte. Basically, the anode and cathode layers of lithium-ion batteries with a liquid electrolyte are powders of a particular material with additives of a binder and if necessary, conductive additives. In such powder materials, liquid electrolyte penetrates deep into the functional layer, and all oxidation-reduction processes occur in its entire volume. The particle size of the material in the powders ranges from fractions of a micron to several microns. In this case, the total thickness of the functional layers can reach dozens of microns. Such a structure of the electrode layers of a lithium-ion battery with a liquid electrolyte allows one to obtain large values of capacity and charge-discharge rate, as well as stable cyclability for hundreds and even thousands of cycles without a notable capacity fading.

The use of solid electrolyte in all-solid-state batteries does not allow one to take full advantage of lithium-ion batteries with a liquid electrolyte. To work the entire volume of the anode or cathode layer, one can use their liquid counterpart, namely, to add a solid electrolyte to the cathode and anode layer. This approach is simply implemented in the so-called pasting (spreading) technologies. As regards PVD methods for deposition functional layers, the task of co-depositing an anode (cathode) material with a solid electrolyte becomes quite difficult. In addition, when applying films by PVD methods, there is a limitation both in the thickness of each individual film and in the thickness of the entire all-solid-state battery. The reason for the impracticality of growing to large thicknesses of each individual layer is the internal stresses in the films obtained by PVD methods. Internal stresses in the films can lead to poor adhesion of the functional layers to the substrate or between dissimilar layers. Thus, an increase in the capacity of all-solid-state lithium-ion batteries due to an increase in its thickness is limited by internal stresses in the films.

3.1.2 Anode and cathode films

The quality of the anode and cathode films of all-solid-state lithium-ion batteries is determined primarily by the best possibility of diffusion of lithium. Thus, dense non-porous films of anodic and cathodic materials allow one to use only the surface layer for lithium diffusion. The entire volume of the film may be out of operation, which negatively affects the battery capacity and charge-discharge rate. Porous films in which lithium diffusion is facilitated due to structural defects, are the most preferred for lithium diffusion. Moreover, for the diffusion of lithium into the film, its entire volume can be used. In addition, porous films have low internal stresses, which is especially important when applying several heterogeneous layers of a solid-state battery with a thickness of more than 1 μm each. As a result, the thickness of the anode and cathode layer is limited due to the diffusion of lithium and internal stresses to 2–3 μm, provided that the necessary porosity is ensured.

Anodic and cathodic films must have sufficient electronic conductivity to provide redox reactions in the cathode and anode of all-solid-state lithium-ion batteries. Pure amorphous silicon has an exceptionally low electronic conductivity. At the same time, the use of silicon composites with a number of metals Al, Ag, W, Zn, Mo and others makes it possible to control the conductivity of silicon composite films even in the presence of oxygen in them. Pure V_2O_5-based cathode films without the addition of lower vanadium oxides are dielectrics, which is why their cyclability is quite poor for the same reasons. A similar situation occurs with dielectric $LiFePO_4$.

3.1.3 Current collector films

The metal films of current collectors should also not be dense, since the internal stresses in them can cause either loss of adhesion of the current collector to the functional layer of the electrode or contribute to the loss of adhesion of the current collector - electrode pair to the solid electrolyte.

3.1.4 Solid electrolyte films

Most of a solid electrolytes should be amorphous after deposition. In the amorphous state, films of solid electrolyte are more resistant to long-term cyclic lithiation-delithiation without changing their conductivity.

3.2 Thermal evaporation method

As applied to the manufacturing technology of all-solid-state lithium-ion batteries, the method of thermal evaporation of materials is effective only for pure metals such as Cu, Al, Ti, Ni and others. In the case of complex compounds containing various atoms, this method is of little use due to the difficulty of obtaining the stoichiometric composition of the final film.

The first experiments on the deposition of metal coatings were carried out by M. Faraday in 1857 with the passage of large electric currents through a metal wire.

The method of thermal evaporation has been well studied and widely used in the industry [1-3]. Figure 3.1 shows the schematic of equipment for the thermal evaporation of materials.

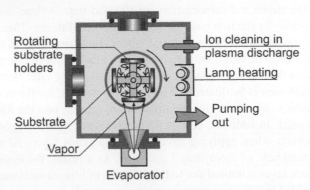

Figure 3.1. Block diagram of the installation of thermal evaporation of materials.

In thermal evaporation installation, the deposition substrates are mounted on a substrate holder. The substrate holder may have various designs with a planetary, cylindrical and flat arrangement of substrates, shafts can also be used to apply materials onto the tape.

The vacuum system consists of two pumps, namely, a low-vacuum (mechanical, oil or oil-free), and a high-vacuum (diffusion, cryogenic or turbomolecular). Preliminary pumping of the working chamber is carried out to a residual pressure of 10^{-3} Pa and above. In all modern installations, there is a system of ionic cleaning of substrates in a discharge plasma before sputtering. If necessary, the substrate is heated to a predetermined temperature with a special heater before sputtering. The evaporator with the deposited material is heated to a temperature above the melting point of the material. When the operating vapor pressure is reached, the shutter opens, and the spraying process takes place.

The process of applying thin-film coatings by thermal evaporation is carried out in several stages: evaporation of the deposited material; transfer of the vaporized substance from the source of the vapor phase to the surface to be coated; condensation of the vaporized material onto the substrate to form a coating.

A necessary condition for obtaining high-quality film coatings is the creation of a high vacuum in the working chamber, which eliminates the oxidation process when the metal is heated to high temperatures and the chemical interaction of vapor phase atoms with residual gas molecules.

Evaporation of a substance takes place when it is heated. When a substance is heated, the kinetic energy of its atoms and molecules increase and become sufficient for them to come off the surface and spread in the environment. With increasing temperature, the energy increases, and the number of atoms detached from the surface increases.

Solids usually melt when heated, and then go into a gaseous state. Some substances pass into a gaseous state, bypassing the liquid phase. This process is called sublimation.

The temperature at which the vapor pressure of a substance above its surface is 1.33 Pa (10^{-2} mm Hg) is called the evaporation temperature of the substance. At this pressure particle mean free path is $\lambda \approx 0.5$ m, which correspond to the characteristic size of the vacuum chamber.

The evaporation rate of a substance is determined by the amount of substance evaporated from a unit area in a unit time, and is expressed by the formula:

$$V_{subl} = 0,0585P_s\sqrt{\frac{M}{T}} \qquad (3.1)$$

where V_{subl} is the evaporation rate, g/(cm²s); P_s is the saturated vapor pressure (1.33 Pa); M is the molecular weight of the vaporized substance, g/mol; T is the evaporation temperature of the substance, K.

In turn, the vapor pressure of a substance depends on the heating temperature. The empirical dependence of vapor pressure on the heating temperature of substances is expressed by the formula:

$$\log P = A - \frac{B}{T} \qquad (3.2)$$

where P is the saturated vapor pressure, mm Hg, T is absolute temperature, K. This equation is valid in the temperature range indicated for each substance. To go from mm Hg to Pa, the value of the constant A should be increased by 2.1249.

According to the data of [4], the constants A and B for the metals of interest are as given in Table 3.1.

Table 3.1. The constants A and B of formula 3.2

Metal	Temperature range, K		A	B
	from	*to*		
Aluminum	724	1279	8,99	15 630
	887	2200	8,23	13 300
Cobalt	1249	2056	9,43	21 960
	1494	3160	9,15	21 400
Copper	970	1083	9,232	17 260
	1083	1290	8,907	16 820
Iron	1094	1535	9,63	20 000
	1310	2850	8,80	18 700
Lithium	325	725	7,50	7480
	439	1330	7,65	7750
Nickel	1157	1455	10,28	21 840
	1371	3000	9,00	19 700
	1455	1884	9,55	20 600
Titanium	1134	1727	8,25	18 640
	1727	3260	9,136	22 110

Data relating to a wide temperature range are intended for rough calculations; the corresponding values are in italics.

Evaporation of materials in a vacuum is carried out using special evaporators. Depending on the source of heating there are:

(1) Resistive evaporation. In this case, the transfer to the gaseous phase occurs under the action of heat released when an electric current passes through a resistive element or an evaporated substance.

(2) Electron beam evaporation. Heating and evaporation of a substance is carried out under the influence of a beam of electrons.

(3) Laser evaporation. The energy source in this method is laser radiation.

(4) Electric arc evaporation. The generation of the gaseous phase occurs due to the burning of an electric arc and the release of heat.

(5) Induction evaporation. The formation of vapors is carried out due to heating when induction currents created by an external high-frequency electromagnetic field pass through a resistive element or evaporated metal.

The evaporator materials must meet the following requirements:

(a) to obtain chemically pure films, the vapor pressure of the evaporator material should be much less than that of the evaporated material;

(b) the surface of the evaporator should be wetted well by the melt of the evaporated substance, which ensures good thermal contact between them;

(c) the evaporated substance must not form chemical compounds with the material of the evaporator.

3.2.1 Resistive evaporation

By design, resistive evaporators are divided into wire, tape and crucible devices. Wire wound, as well as tape evaporators, are made of refractory materials W, Mo, Ta. Crucible evaporators are made of C, BN, TiB_2-BN, ThO_2, Al_2O_3, Mo. Table 3.2 gives recommendations on the use of materials of evaporators for thermal evaporation of metals used in the manufacture of all-solid-state lithium-ion batteries.

Table 3.2. Recommended Evaporator Materials

Evaporated material	*Recommended Evaporator Materials*	
	Wire, tape	*Crucible*
Al	W	C, BN, TiB_2-BN
Cu	W, Mo, Ta	Mo, C, Al_2O_3
Ni	W	
Ti	W, Ta	C, ThO_2
Li	Mo	

The advantages of the method of resistive evaporation of materials for all-solid-state batteries include the simplicity of the method and the possibility of applying metal films in a wide range of rates. Coating on tapes up to several meters long allows for high productivity of the metal film deposition process.

The disadvantages of the method include: the difficulty of controlling the flow of atoms; low ionization of atoms during evaporation (less than 1%) and, as a result, low adhesion of the resulting films to the substrate material.

3.2.2 Electron beam evaporation

Electron beam heating is the most widespread method of vacuum deposition of films. There are many different schemes of evaporators and installations for vacuum deposition using electron beam heating.

Figure 3.2 shows the most common electron beam heating scheme.

A high voltage is supplied to the cathode and anode of an electron gun from the power source. Depending on the type of gun, it can range from 5 to 60 kV. The thermal cathode (direct-heated or indirectly-heated) in the process of thermionic emission emits electrons, which are accelerated in the direction of the water-cooled anode. The deflecting and focusing electromagnetic systems (coils) direct the formed beam to the evaporated material in the form of a rod or a water-cooled crucible. Electrons under the influence of an electric field accelerate and acquire significant kinetic energy. An electronic bombardment on the surface of the sputtered material occurs. The main part of the kinetic energy of electrons is released in the form of heat and expended on the evaporation of the material. However, a significant part of it is lost on the formation of secondary electrons (15-30%) and X-ray radiation (about 0.1%). The energy of reflected electrons is spent on heating the walls of the chamber, the substrate and other devices in the chamber.

The most significant parameter in electron beam deposition of coatings is the power of the beam. As with other methods of film deposition by thermal evaporation,

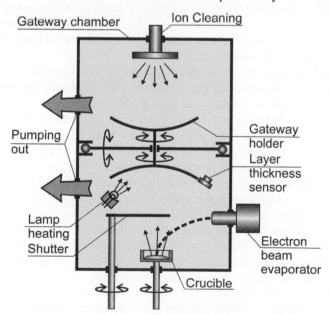

Figure 3.2. The structural scheme of the installation of electron beam evaporation of materials.

the energy of the evaporated atoms in the vapor stream is small and amounts to 0.2–0.3 eV; the degree of ionization of the particles is 0.01–1.2%. However, this method is characterized by a high density of the vapor stream, which ensures high performance of the thermal deposition process. The advantages of the electron beam evaporation method include the ability to work at lower pressures (up to 10^{-2} Pa and lower) than with resistive evaporation. Such pressure allows free diffusion of atoms of the substance to the substrate without collision with residual gas molecules in the working chamber. The water-cooled crucible usually does not introduce pollution during the sputtering process, therefore high purity of the deposited metals is achieved. The disadvantages are the increased complexity of the installations during their manufacture and operation, the presence of hard radiation at high accelerating voltages. The method of electron beam deposition is especially suitable when coating large surfaces, for example, on a tape in continuous installations [5].

3.2.3   Other methods of thermal evaporation

Other PVD thermal methods for depositing functional layers of all-solid-state lithium-ion batteries will also be briefly described. These are (*i*) laser ablation and (*ii*) Arc-PVD methods. The disadvantages of these methods make them hardly suitable for the manufacture of all-solid-state batteries.

In the laser ablation method, the target is irradiated with pulsed UV radiation from an excimer (KrF, CO_2) or Nd: YAG laser. The radiation intensity is 10^8–10^9 W/cm^2, the duration is several nanoseconds, which is sufficient for ablation of a substance (metals, metal oxides, etc.) at the target heating point.

In a generalized form, the advantages of the method include: high quality of deposited films; a high degree of compliance of cationic stoichiometry of formed films, to the composition of the target material; high morphological uniformity of the formed film; very high sputtering rate; almost complete absence of film contamination by the components of the camera materials.

The disadvantages of the method include the small geometric size of the homogeneous deposition zone during ablation in vacuum, due to the small diameter of the sputter zone of the products, as well as the possibility of contamination of the film with solid particles and drops of the melt of the target material at high deposition rates. This is the main drawback that makes this method unsuitable for the manufacture of all-solid-state lithium-ion batteries with a large area.

In the method of electric arc evaporation, a pressure of about 10^3 Pa is created in the sputtering chamber. Between the anode, ignition electrode and the cathode made of the deposited material, a voltage is applied. The ignition electrode is used to ignite an electric arc. This action is performed by briefly touching the ignition electrode of the cathode surface. The local temperature of the cathode spot is extremely high (about 15,000 °C), which causes intense evaporation and ionization of the cathode material in them and the formation of high-speed (up to 10 km/s) plasma flows propagating from the cathode spot into the environment.

The advantage of the vacuum-arc evaporation method consists of the following. The resulting coatings have high adhesion to the substrate, a dispersed structure and low porosity. With this method, it is possible to sputter metals and complex alloys with a high rate. Systems based on vacuum-arc evaporation of materials provide deposition rates up to units of microns per minute. To obtain multi component coatings, several cathodes made of various materials can be used simultaneously.

For deposition of coating of carbides, nitrides or metal oxides, reactive gas is supplied to the chamber.

The disadvantage of the vacuum-arc evaporation process is that if the cathode spot remains at the evaporation point for too long, it emits a large amount of particulate or droplet phase.

As it was noted at the beginning of the chapter, one of the main requirements for the functional layers of all-solid-state lithium-ion battery is the creation of sufficiently porous films with a large number of structural imperfections. In the method of electric arc evaporation, dense films having low porosity are obtained. In addition, the presence of a droplet phase in films can adversely affect the quality of the functional layer of all-solid-state lithium-ion battery.

3.3 Magnetron sputtering method

Magnetron sputtering systems are diode-type sputtering ones in which the sputtering of the target material occurs due to the bombardment of the target surface with working gas ions. Ionization of the working gas occurs in the plasma of an abnormal glow discharge. A high sputtering rate in such systems is achieved by increasing the ion flux density due to the localization of the plasma near the sputtered target surface using a strong transverse magnetic field.

There are several types of sputter systems for thin film deposition. Among these sputter systems, the main model is a DC diode system. The principle of operation of the DC magnetron sputtering system is shown in Fig. 3.3.

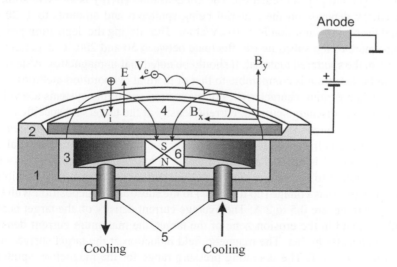

Figure 3.3. The scheme of operation of the magnetron sputtering system: 1 - magnetron case; 2 - flange-base of target; 3 - magnetic circuit; 4 - target-cathode; 5 - cooling pipes; 6 - permanent magnet; S, N - magnet poles; V_e is the electron velocity; V_i is the ion velocity; B_x, B_y - projections of the magnetic field induction vector; E is the electric field.

The main elements of the device are the cathode-target, anode and the magnetic system. The field lines of the magnetic field are closed between the poles of the magnetic system. When a constant voltage is applied between a target having a negative potential and an anode having a positive potential, an inhomogeneous electric field emerges. An anomalous glow discharge appears in the anode-cathode gap. The presence of a closed magnetic field at the target surface allows localization of the discharge plasma directly at the target. Electrons emitted from the cathode are captured by a magnetic field and undergo a cycloidal motion along closed trajectories at the target surface under the action of the Lorentz force. The electrons move as if they are trapped. A magnetic field returns them to the cathode, while an electric field repels them from the surface of the target. Electrons cycle in this trap, colliding with the atoms of the working gas and ionizing them. Most of the energy of an electron, before it hits the anode, is spent on ionizing the working gas. An increase in ionization leads to an increase in the concentration of positive ions that bombard the surface of the target. Thus, an increase in the intensity of ion bombardment leads to an increase in the sputtering rate and, accordingly, an increase in the deposition rate of the film.

The discharge plasma is concentrated only in the region of the magnetic trap near the target, and its shape is determined by the geometry and magnitude of the magnetic field.

The magnetron discharge has a number of specific features. In this discharge, there is no bombardment of the substrate by high-energy secondary electrons, leading to additional heating of the substrate during the deposition process. The sources of substrate heating in this process are the condensation energy of sputtered atoms, the kinetic energy of the deposited atoms, the energy of neutral ions reflected from the target and plasma radiation. The condensation energy is 3–9 eV/atom, the kinetic energy depends on the material being sputtered and amounts to 1–20 eV/atom and the plasma radiation is 2–10 eV/atom. Thus during the deposition process, the temperature of the substrate can fluctuate between 50 and 200 °C depending on the nature of the sputtered material. It should be noted that in magnetron systems the heating of the substrate is comparable to that of thermal evaporation methods.

The main operating characteristics of magnetron sputtering systems are voltage at the electrodes, discharge current, current density at the target and specific power, magnitude of the magnetic field induction and working pressure. The voltage range on the electrodes is from 100 to 800 V. Usually the anode is at zero potential, and a negative potential is applied to the target. In some cases, a positive bias of up to 50 V is applied to the anode, and in some sputtering systems it is possible to supply a negative bias voltage (up to 100 V) to the substrate for deposition with bias. Operating currents are 0.5 to 2 A. The average current density on the target is 50 to 300 mA/cm², and in the erosion zone of the target, the maximum current densities can be significantly higher. The magnetic field induction at the target surface has a value of 0.03 to 0.1 T. The operating pressure range for the magnetron sputtering system is from 0.01 to 10 Pa.

The current–voltage characteristic of a planar magnetron system can be described by an equation of the form

$$I = KU^n \quad (3.3)$$

where I is the discharge current; U is the operating voltage; K is the coefficient of proportionality; $n = (5–9)$ is an exponent that depends on the efficiency of electron capture by a magnetic trap [6].

The current-voltage characteristics over a wide range of pressure changes and magnetic field induction are linear on a logarithmic scale. Significant deviations from linearity are observed at low pressures and small magnetic field inductions.

Magnetron sputtering can be carried out both on direct and alternating current (RF magnetron sputtering). In RF magnetron sputtering, high-resistance semiconductor and dielectric materials are easily sprayed. The main advantages of the magnetron sputtering method are the high film deposition rate and reproducibility of the composition of the sprayed material. Compared to thermal deposition methods, the magnetron method gives a higher level of adhesion of the resulting films.

The method of magnetron sputtering in the manufacture of all-solid-state lithium-ion batteries is the most universal of all PVD film deposition methods. This method in both DC and RF versions can be applied to almost all materials, which are necessary in the manufacture of most cathode materials and solid electrolytes. In addition, this method is suitable for reactive magnetron deposition when chemical interaction of the sputtered material with the working gas ions is necessary or the reactive gas is mixed with the working gas.

The universality of the magnetron sputtering method allows the deposition of functional layers of all-solid-state battery in one unit. The magnetron sputtering system can be equipped with several DC and RF magnetrons, as well as combining magnetrons with thermal evaporators. Structural diagrams of such equipment are shown in Figs. 3.4 and 3.5.

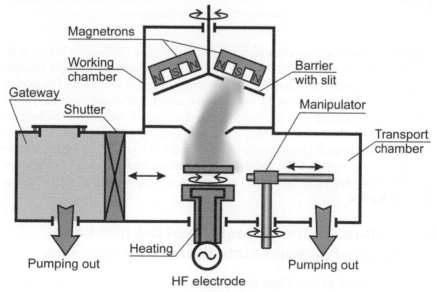

Figure 3.4. The structural diagram of a magnetron sputtering unit with two magnetrons

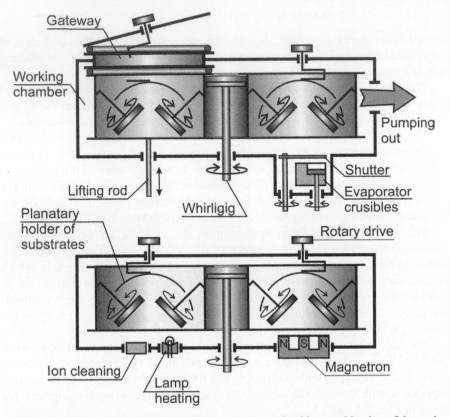

Figure 3.5. The structural diagram of the sputtering unit with a combination of thermal and magnetron methods of film deposition

An ideal option for the manufacture of all-solid-state lithium-ion batteries is a cluster-type installation in which several units are combined by a single transport system for substrates. For example, unit of thermal or magnetron sputtering of metal current collectors, unit of DC or RF magnetron sputtering of anode materials, unit of RF magnetron sputtering of solid electrolyte and unit of RF magnetron sputtering of cathode materials. In a cluster system for manufacturing all-solid-state lithium-ion batteries, contact with the atmosphere of the functional layers of lithium-containing materials, namely lithium, solid electrolyte and cathode materials can be eliminated. As a result of passing all technological operations, a completely ready all-solid-state battery is obtained.

3.4   Features of the deposition of the functional layers of all-solid-state lithium-ion batteries

Of the three stages of thin films manufacturing by PVD methods, the stages of transfer of a substance from a source of a sprayed material (an evaporator or a target) to a substrate were considered. The next stage in the formation of the film is the condensation of the evaporated or sputtered material on the substrate. The structure of the resulting film is influenced by several factors that are somehow related to the energy of the condensed particles. In the resistive and electron beam methods of thermal evaporation, energies of particles hitting the substrate, is small and amounts to 1 eV. In the method of magnetron sputtering, this value is 10 to 20 eV. The second energy factor affecting the process of condensation of the evaporated material is the temperature of the substrate. The third factor affecting the structure of the film is the pressure in the working chamber. The influence of these factors is widely covered in literature.

Once more factor affecting the structure of the deposited film and its internal stresses is the quality of the substrate surface. These can be seen by examples of substrates of single-crystal silicon and titanium foil how the structure of the deposited films depends on the preparation of their surfaces. Before sputtering the substrate, as a rule, it is subjected to ion cleaning to remove surface contaminants. If the surface is smooth, the thin surface layer of the substrate removed by ion cleaning does not change its roughness. The structure of the deposited films on such a surface is uniform and quite dense. In Fig. 3.6, examples of the structure of Si-Al composite films on a silicon substrate (a) and Ti foil (b) obtained after ion bombardment of the substrate are given.

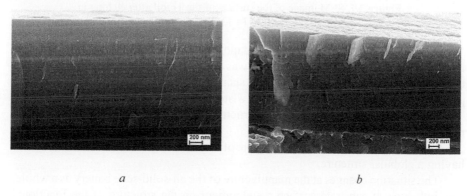

a   *b*

Figure 3.6. The structure of Si-Al films on (a) silicon substrate and (b) Ti foil after ion treatment of the substrates

The development of the substrate surface, for example, of a titanium foil by etching it in an $HF : H_2SO_4 : H_2O = 1 : 1 : 20$ solution for 40 seconds, is shown in Fig. 3.7.

On the developed surface, the structure of the deposited films has the shape of globules. Figures 3.8 (a-d) show SEM images of the cleavage and surface of Si-Al and V-O films deposited on a titanium foil with the surface shown in Fig. 3.7.

Numerous experiments on the deposition of Si-Al and V-O films on a Ti foil with a developed surface at different substrate temperatures, power and pressure in the working chamber did not reveal a strong dependence of the film structure on these parameters. The substrate heating varied from ambient temperature to 200 °C. The power of the magnetron discharge varied from 100 to 700 W. The argon pressure in the working chamber varied from 0.2 to 17 Pa.

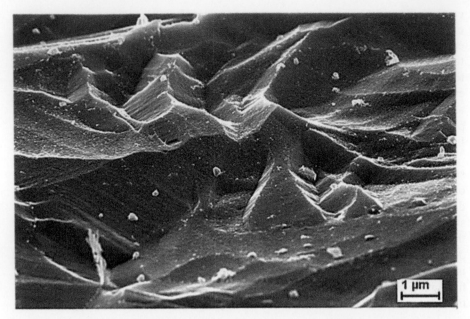

Figure 3.7. SEM image of the surface of the Ti foil after etching in
a HF : H_2SO_4 : H_2O = 1 : 1 : 20 solution for 40s

It should also be noted that films deposited on a developed surface had low internal stresses. The absence of stresses was confirmed by the fact that Ti foil 10 μm thick with a deposited film retained a flat shape. While the film on the Ti foil without preliminary development of the surface twisted the foil with the film outwards.

Thus, the main factor affecting the structure of the deposited film was the surface roughness of the substrate.

It is worth noting that the diffusion of lithium is greatly facilitated in films having a globular structure.

The situation changes at the manufacture of the all-solid-state battery as a whole. The influence of substrate surface development on the structure of the functional layers decreases from layer to layer. The deposition of several heterogeneous layers with thicknesses of several hundred nm gradually planarizes the surface. In Fig. 3.9 the structure of all-solid-state battery on a Ti foil with $LiCoO_2$-LiPON-(Si-Al) layers is shown.

In this structure, the $LiCoO_2$ layer lying directly on the Ti foil has a porous structure. The LiPON layer lying on $LiCoO_2$ is dense with no signs of porosity. Accordingly, a Si-Al layer deposited on a planarized surface no longer shows signs of porosity. As already mentioned, the upper Si-Al layer with a dense structure is not optimal for lithium insertion and extraction. In addition, internal stresses appear in such a structure, which can lead to peeling of the upper Si-Al layer from the LiPON layer.

Thus, the development of the substrate surface is the main factor affecting the structure and internal stresses of the functional layers of all-solid-state lithium-ion batteries.

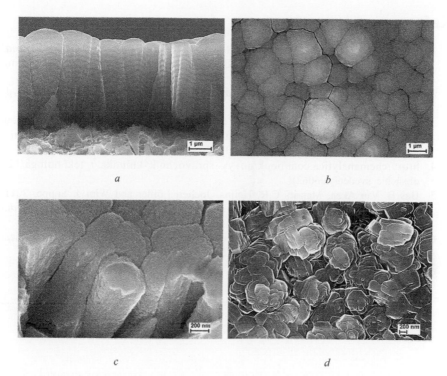

a   *b*

c   *d*

Figure 3.8. SEM images of the surface and cleavage of Si-Al films (*a, b*) and V-O films after deposition (*c*) and after high-temperature treatment at a temperature of 400 °C for 80 minutes in oxygen (*d*) deposited on a titanium foil with a developed surface

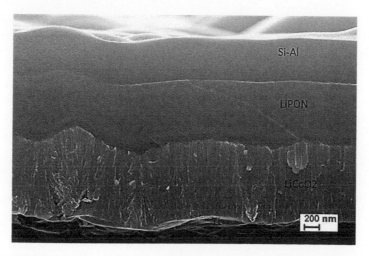

Figure 3.9. SEM image of all-solid-state battery structure Ti-LiCoO$_2$-LiPON-(Si-Al)

References

1. Makhlouf, A.S.H. 2011. *Current and Advanced Coating Technologies for Industrial Applications*. Max Planck Institute of Colloids and Interfaces. Woodhead Publishing Limited, Germany.
2. https://www.grandviewresearch.com/industry-analysis/physical-vapor-deposition-pvd-market.
3. Shahidi1, Sheila, Bahareh Moazzenchi and Mahmood Ghoranneviss. 2015, A review – Application of physical vapor deposition (PVD) and related methods in the textile industry. *Eur. Phys. J. Appl. Phys.* 71: 31302.
4. http://chemanalytica.com/book/novyy_spravochnik_khimika_i_tekhnologa/12_obshchie_svedeniya/6062)/
5. Bas B. Van Aken, Maurits C.R. Heijna, Jochen Löffler and Wim J. Soppe, 2011. Dynamically deposited thin-film silicon solar cells on imprinted foil. *Energy Procedia* 10: 88–93.
6. Thornton, J.A. 1978. Magnetron sputtering: Basic physics and application to cylindrical magnetrons. *J. Vac. Sci. Technol.* 15: 171–7.

Diagnostics of Functional Layers of All-Solid-State Thin-Film Lithium-Ion Batteries

4.1 Scanning electron microscopy

4.1.1 Scanning electron microscopy method

Electron microscopy is one of the most informative methods for characterization of the functional layers of all-solid-state thin-film lithium-ion batteries. This is primarily due to the typical size of the layers and their structural features, inherent in PVD technologies. The most common method for studying the morphology of battery materials is Scanning Electron Microscopy (SEM). It differs from optical microscopy in that it allows to work in a wide range of magnifications – from 50 times to 150,000 times, which in approximately 150 times exceeds the magnification limit of the best optical microscopes. The limited possibilities of optical microscopy are due to the relatively long wavelength of light illuminating the object. The wavelength of most short-wave radiation – ultraviolet lies in the range from 10 to 400 nm, i.e., outside the visible region of the spectrum, which, depending on the individual characteristics of the observer, can range from 380–400 nm to 760–780 nm. According to the Abbe formula, the resolution of the optical system (for simplicity, the resolution can be considered as the minimum size of the details to be distinguished) in the limiting case is 0.61 of the wavelengths, which for the optical region of the spectrum gives a value of 200 nm at best. To overcome this limitation, various methods are used – ultraviolet microscopy, near-field microscopy, etc. For ultraviolet microscopy, the wavelength can reach 50 nm, and the resolution, respectively, can be estimated at about 30 nm. But even these 'short' ultraviolet waves do not 'see' the morphological features of an object with characteristic sizes less than a micron. Namely, they are of most interest to Solid-State Lithium-Ion Batteries (SSLIBs) developers, because of containing information about the thickness of the functional layer, its morphology, crystalline structure, porosity and even the adhesion of the layers to the substrate.

In order for the radiation to scatter on the inhomogeneities of the sample of submicron size, its wavelength must be less than the size of the inhomogeneity itself. As such radiation the waves (or rather wave packets) of probability density, which are called de Broglie waves, can be used. According to de Broglie's hypothesis, the wavelength is related to the particle momentum by the relation

$$\lambda = h/p, \tag{4.1}$$

where h is the Planck constant and p is the particle momentum. In the case when the particle's velocity is much less than the velocity of light, the relation $p = mv$ or $p = \sqrt{2mE}$ is valid, and Equation (4.1) can be represented in the form $\lambda = h / \sqrt{2mE}$, where m is the mass of the particle and E is its kinetic energy. If a particle whose energy is initially negligible is accelerated by a potential difference $\Delta\varphi$, then the expression for the wavelength of such a particle has the form

$$\lambda = \frac{12.25}{\sqrt{\Delta\varphi}} \text{ Å}. \tag{4.2}$$

It can be seen that for the accelerating potential of 15,000 V, the de Broglie wavelength is 0.1 Å. The packet of such waves is much smaller than the morphological features of the object under study and will be effectively scattered by them.

In addition, charged particles have other advantages, including relatively simple controllability using electric and magnetic fields. For electron and ion microscopy, this means the possibility of focusing, changing the beam current density (fluence), scanning the beam and adjusting the shape of the beam to obtain a circular cross section. In a scanning electron microscope, a converging beam of monoenergetic electrons with the required current density is focused on a sample located at a working distance from an edge of an objective lens. The scanning system allows one to move the beam along the surface of the sample, within the so-called raster. At each moment of time, one of the electron detectors registers the intensity of the scattered electrons (signal). Combining the detector signal with the coordinates of the beam one can get a kind of photographic image of the surface of the sample.

In real instruments, the resolution of the microscope is affected by spherical and chromatic aberrations, as well as astigmatism of the electron beam. The nature of such distortions lies in the imperfection of electromagnetic and electrostatic lenses (the presence of residual magnetization, the impossibility of manufacturing absolutely symmetric or identical parts of an electron column), in the final size of the source, and the impossibility of providing an ideal vacuum, temperature, acoustic and other operating conditions for an instrument that operates in the mode of relatively frequent change of samples. This set of factors limits the passport resolution of devices with a thermionic source at a level of 3 nm, and devices with a thermal field emission source at a level of 1 nm.

4.1.2 Design and principle of operation of a scanning electron microscope

The operation of SEM is based on the above described principle when an electron beam is used as an electron probe. The term 'scanning' means that, unlike an optical microscope, where an entire object is irradiated with scattered light, in the SEM the test sample is scanned by a focused beam that probes the surface, scattering electrons, which are gathered by detectors reading information about the surface.

A scanning electron microscope is a high-vacuum device, since at low vacuum the electron beam is strongly scattered and absorbed, which makes it impossible to focus it. Therefore, the working vacuum in the chamber of a standard microscope should be 10^{-5} Torr, or higher. Modern instruments designed to study a wide class of objects, including biological objects, allows one to work in low vacuum mode (up to 1 Torr) and in the so-called natural environment mode (up to 20 Torr). The main structural elements of the SEM (Fig. 4.1) are: a column, vacuum system, a detection system and control and visualization units.

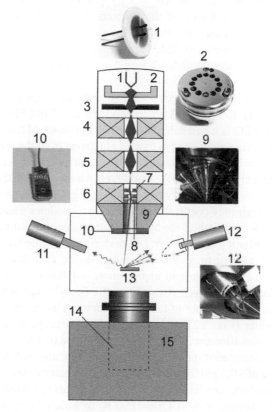

Figure 4.1. Scheme of a scanning electron microscope 1 - cathode, 2 - Wehnelt cylinder, 3 - anode, 4-6 - condenser lenses, 7 - deflecting coils, 8 - reflected electron detector (inlens secondary electron (SE) detector), 9 - polepiece, 10 - 4 sectional scattered electrons detector, 11 - XRI detector, 12 - Everhart-Thornley Secondary Electron detector (ETSE), 13 - sample, 14 - turbomolecular pump, 15 - vacuum system

The primary electron beam is formed by an electron gun located in a vacuum column, which is a vertically mounted cylindrical vacuum chamber. As an example, a microscope containing: an electron gun, consisting of a cathode 1 and an accelerating system 2, 3; a focusing system 4-6; a deflecting system 7 and a system of axial detectors 8, 10 could be considered. The vacuum column rests on the sample chamber in which sample 13, the X-ray detector 11 and Everhart-Thornley detector of the secondary electrons 12 are located. The working pressure in the vacuum chamber is achieved by using the turbomolecular 14 and foreline pump. In certain systems (for thermal field and ion sources), ion-absorption pumps are used in the column area, which, after preliminary pumping, create a pressure of 10^{-7} or less Torr.

Two types of cathodes (emitters) are used in electron guns: with thermionic emission and field emission. The first type known as Thermionic Guns, are devices with thermionic emitters. The emitter of such a gun (Fig. 4.1) consists of a V-shaped tungsten filament with a diameter of about 200 microns, to which a small voltage is applied. This voltage heats the filament to a temperature of 2000-2700 K, at which thermionic emission from the tip of the filament reaches the value necessary for normal SEM operation. Electrons emitted from cathode 1 are accelerated by an electric field of voltage from 1 to 30 kV applied between the cathode and anode 3. An electrode called Wehnelt cylinder 2 serves to pre-focus the electron beam and control its intensity.

The second type includes electron guns with field emission, called Field Emission-Guns (FEG). In turn, FEG are subdivided into Cold FEG or simply FEG. The action of Cold FEG is based solely on the phenomenon of field emission, when only an external electric field is used to extract electrons from the metal. The cathode metal in vacuum can be considered as a rectangular potential well in which free electrons are located at and above the Fermi level. An external field makes the width of the potential barrier at the metal boundary finite, giving it a saw-toothed shape. As a result, the transparency coefficient of the barrier becomes greater than zero and higher, the smaller is the width of the barrier. In this case, part of the electrons goes beyond the metal without loss of energy, i.e., makes the so-called quantum tunneling. In devices simply referred to as FEG, to increase the emission of electrons, the cathode is heated to a temperature of approximately 1,800 K. In this case, a part of the electrons in the potential well rises above the Fermi level, where the potential barrier is narrower so the number of tunneling electrons increases.

The FEG emitter (Fig. 4.2) is a rod 5 made of a tungsten single crystal and mounted on a tungsten filament 2. The rod has the shape of a needle with a diameter of several tens of micrometers, tapering to the lower end, and ending with a tip with a radius of curvature of ~0.5 μm. Approximately in the center of the rod there is a bulb 6 of zirconia. In working conditions, a current passes through the tungsten filament, heating the rod to a temperature of 1700-1800 K, at which the zirconia drains along the rod in the direction of the edge. The zirconia layer reduces the electron work function, and the extractor 4 lowers the electric potential around the tungsten rod, so that the potential barrier at the tungsten boundary turns from infinite into a barrier of finite width. Thus, the cathode emits mainly 'cold' electrons, which leave the cathode by quantum tunneling. This phenomenon is also called 'field emission' or 'cold emission', although the temperature of the rod is rather high. The small radius of the rounding R, with which the surface charge density σ is related as $\sigma \sim 1/R$, creates a high charge density at the tip. The shape of the electrode practically does not affect the transparency of the potential barrier, but a high electron density increases the number of electrons that tunnel through the barrier.

The principle of image acquisition by SEM is as follows. The electrons, emitted by the electron gun, are accelerated by the anode and travel through the column by inertia, where the electron beam is centered by deflecting coils, its peripheral part is

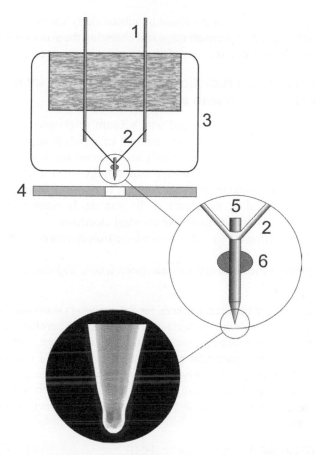

Figure 4.2. Scheme of field emission gun: 1 - current lead, 2 - tungsten filament,
3 - suppressor (–500 V), 4 - extractor (from +4 kV to +5kV), 5 - tungsten rod,
6 - zirconium oxide

cut off by limiting apertures and its asymmetry is eliminated by a stigmator and is finally focused at the spot with diameter of the order of 1 nm in the case of a FEG and of the order of 3-4 nm for the thermionic gun. This beam is deflected by scanning coils and move line by line, point by point, scanning the sample. When an electron beam hits any point of the sample, secondary electrons are emitted from the sample material. Secondary electrons are captured by a detector, which converts the electron energy into an electrical signal, the value of which is proportional to the number of electrons gathered by a detector. The larger the fraction of secondary electrons at a given point in the sample, the greater is the amplitude of the signal recorded in one measurement event. The movement of the electronic probe within the raster occurs synchronously with the plotting image on the monitor. If the brightness of the corresponding pixel on the monitor changes in proportion to the amplitude of the detected signal at each point on the surface of the sample, one can get an image of the surface on the screen. However this will not be an image in the visible range of electromagnetic waves, but in the range of substantially shorter de Broglie waves. By turning and tilting the specimen stage and changing the size of the raster, one can get an image of the required sections of the sample with the necessary magnification.

4.1.3   Particles and radiation generated by the interaction of an electron probe with a sample

When a sample surface is bombarded with a beam of primary electrons (electron probe), the surface emits electrons, the amount of which is approximately 30% of the number of primary electrons. In literary and Internet sources, these electrons are often called secondary. This confuses the terminology, since among these electrons there are both truly secondary ones that belong to the sample and electrons of the primary beam scattered by the atoms of the sample. In electron microscopy, the following terminology is used to denote detected electrons:

- Secondary Electrons (SE) – electrons whose trajectories cross the surface of the sample once;
- Back Scattered Electrons (BSE) – electrons whose trajectories cross the sample surface twice [1].

However there are no direct methods for determining the number of intersections of the electron trajectory with the surface of the sample. Therefore, the division into secondary and backscattered electrons is conditional in nature, and the criterion for assigning electrons to a particular group is their energy. The electron energies lie in the range of values $[0, E_0]$, where E_0 is the energy of the primary beam. The energy distribution of electrons emitted by surface is illustrated in Fig. 4.3, in which the entire energy interval is divided into two regions, namely backscattered and secondary electrons.

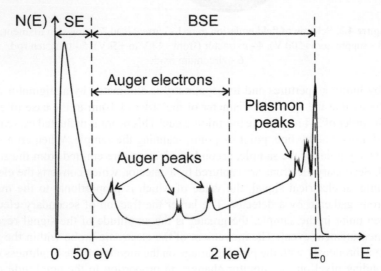

Figure 4.3. Qualitative dependence of the number of emitted electrons on their energy. E_0 - energy of primary electrons

Electron backscattering can occur both elastically and inelastically. Elastic scattering occurs as a result of collisions of high-energy electrons with atomic nuclei partially shielded by electron shells. In elastic scattering, the direction of the electron velocity vector changes, while its magnitude and kinetic energy of the electron remain unchanged. Such scattering is also called Rutherford because it is described by Rutherford's formula

$$d\sigma = \left(\frac{Ze^2}{4E}\right)^2 \frac{d\Omega}{\sin^4(\Theta/2)}, \tag{4.3}$$

where $d\sigma$ is the differential scattering cross-section, Z is the atomic number of the nucleus, e is the electron charge, E is the electron energy, Θ is the two-dimensional scattering angle, Ω is the solid angle into which the electron is scattered. It can be seen from formula (4.3) that the differential scattering cross-section, and, consequently, the number of elastically scattered electrons, grows as Z^2. The angle of deviation from the direction of incidence during elastic scattering can vary in the range from $0°$ up to $180°$, but it's most probable value is in the order of units of degrees. Most electrons undergo several scattering events, as a result of which the scattering angle can take on any value.

In many cases, a closed nucleus-electron system can be considered conservative and the laws of conservation of energy and angular momentum must be satisfied in it. In real conditions, the nucleus-electron system is an open system for which conservation laws are not satisfied. Therefore, some of the energy of the primary electron is transferred to the entire sample, i.e., scattering is quasi-elastic. But the amount of energy transferred to the sample during each event of scattering, treated as elastic, is about 1 eV, which is negligible compared to the initial electron energy in the beam (1-30 keV). The elastic and quasi-elastic scattered electrons on the graph in Fig. 4.3 form a sharp peak in the vicinity of the point E_0.

Besides elastic or quasi-elastic scattering by nuclei, some of the electrons of the primary beam can undergo inelastic scattering by free electrons of metals and by valence electrons of metals and dielectrics. When scattered by free electrons, in an electron gas density oscillations are excited, called plasmons, to which part of the primary electron energy is transferred. When scattering by free electrons, true secondary electrons can be knocked out. Part of the primary electron energy in this case will be spent on overcoming the potential barrier at the metal-vacuum interface and on the residual kinetic energy of the secondary electron in vacuum. Inelastically scattered electrons are those where the energy lies below the peak E_0 but exceeds 50 eV.

The number of inelastically scattered electrons in Fig. 4.3 with decreasing energy tends to be zero. However, a rather sharp peak of low-energy electrons is observed in the experimental curve. These are the so-called true secondary electrons [2]. They are formed as a result of inelastic scattering of primary electrons by free or weakly bound electrons of the outer shells. If such shell electron as a result of a collision acquires energy sufficient to overcome a potential barrier at the substance–vacuum interface, it can leave the sample, but its energy will be very small – in the order of 50 eV and lower. In practice, secondary electrons include all electrons with energies less than 50 eV. Although part of the inelastically scattered electrons also has energy below 50 eV, their share in the total number of secondary electrons is negligible.

Primary electrons, if possessing sufficient energy, can knock out secondary electrons from the inner atomic shells, leading to ionization of the atoms. Since such an ionized state of an atom is unstable, the electronic subsystem seeks to minimize energy by filling a vacancy with an electron from one of the overlying energy levels. On transition to the underlying level, the electron energy can be released in the form of a quantum of Characteristic X-ray Radiation (CXR), or transmitted to a third electron, which is forced to leave the atom. The second process was known by the name of its discoverer Pierre Auger – the 'Auger effect', and the electron released in this case, to which the excess energy was transferred, is called the 'Auger electron'. The energy of the Auger electron does not depend on the energy of the exciting radiation but is determined by the structure of the energy levels of the atom. Therefore, the spectrum of Auger electrons is discrete and carries information about the internal structure of the atom.

In addition to the above processes of interaction with the sample, primary electrons during braking in the field of the sample atoms generate continuous X-ray radiation. Since energy losses during braking are determined by an arbitrary configuration of the electron-atom system, the spectrum of braking X-rays radiation is continuous. In the scattering of high-energy electrons by the nuclei of the atoms of the sample, the latter shift relative to the equilibrium positions, as a result of which vibrations of the crystal lattice appear. In this case, it can be said that primary electrons generate phonons - quasiparticles, whose energy is equal to the minimum excitation energy of the normal vibrational mode of the crystal lattice. Phonon generation, as well as an increase in the energy of the chaotic motion of free electrons in metals, leads to a significant heating of the sample. Therefore, when studying materials for lithium-ion batteries by SEM methods, it is necessary to ensure good thermal contact of the sample with the microscope stage.

All these types of generations of secondary electrons, radiation and quasiparticles play the role of response signals that carry information about the topography, morphology and elemental composition of the sample. To form an image of the surface of the sample on the monitor of SEM, secondary electrons, Auger electrons and CXR are used. To obtain an image, detectors of various sensitivity and geometry can be used, which, depending on the task, enables in procuring various types of surface contrasts. In addition, there is the possibility of combining frames acquired by using various detectors.

4.1.4   Secondary electron detection

Depending on the model and configuration, the SEM may have a different number of secondary electron detectors. As a rule, modern microscopes as a standard are equipped with three main detectors: an inlens backscattered electron detector, a four-segment solid-state detector of backscattered electrons, located directly on the pole piece and Everhart-Thornley detector of a true secondary electrons located in the sample chamber.

Inlens detector

Almost all modern scanning microscopes have a secondary electron detector located inside the objective lens. Such a detector is called inlens SE and works as follows. With a horizontal arrangement of the sample, most of the primary electrons scatter in the opposite direction. The braking field of the electron gun (beam booster) collects the backscattered electrons and accelerates them inside the last condenser lens. Accelerated backscattered electrons hit the scintillator of a coaxial inlens electron detector, generating photons. Through the optical fiber, these photons from the microscope column enter the photomultiplier, where, on reaching the photocathode, secondary electrons are knocked out. Secondary electrons are accelerated by a cascade of electrodes (dynodes) of the photomultiplier, the secondary electron emission coefficient of which exceeds 1. Thus the electron flux increases avalanche-like from dynode to dynode, as a result, the current in the anode circuit of the photomultiplier exceeds the initial photocurrent 10^5 times.

The inlens detector is most convenient for an operation when the sample is located close to the lens. When operating in high vacuum, an intra-lens detector allows for extremely high contrast images of the surface. If the surface of the sample is perpendicular to the beam of primary electrons, then with the help of an intra-lens detector it is possible to study structures with a developed relief, for example, to 'look' into deep grooves or pits that are obscured for other detectors.

Solid-state detector of backscattered electrons

Solid-State Detector of BackScattered Electrons (SSD-BSE), which due to its location are often referred to as an onlens detector, is a four-segment semiconductor detector mounted directly underneath the pole piece (Fig. 4.4), that ensures maximum detection efficiency. If a detector is not necessary, it can be removed and placed in a special holder in the sample chamber. Since the detector is mounted under the last lens, i.e., in the immediate vicinity of the sample, this leads to some restrictions in the range of tilt angles when working with an eucentric specimen stage. An advantage over the inlens detector is the large area of the detector and the possibility of obtaining two types of contrast.

The detector consists of four segments, formed by semiconductor diodes (Fig. 4.5), and is closed by a metal grid, to which a positive potential is applied. The diodes, described in Fig. 4.4 as A, B, C, D are connected to the input of the operational amplifier. It is notable that all four segments can be connected independently from each other and in different combinations. For the photodiode two connection modes are possible: photovoltaic and photoconductive mode. In this case, the photoconductive mode is used, as photodiodes operating in the photoconductive mode have greater sensitivity, lower equivalent capacitance (due to the large width of the depletion layer) and, therefore, more small pulse rise time than in photovoltaic regime. The reverse bias applys to the diode in the photoconductive regime forming a depletion layer in the region of p-n junction. When a high-energy electron strikes the diode, it generates electron-hole pairs in the diode bulk. If these pairs appear in the depletion region or one diffusion length away from it, they are separated by an

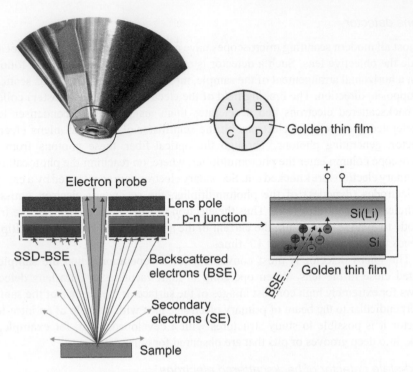

Figure 4.4. Solid-state detector of backscattered electrons (SSD-BSE) based on p-n junction

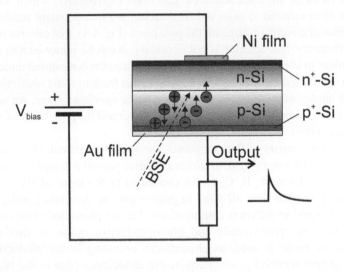

Figure 4.5. Scheme of connection of a solid-state BSE-detector in photoconductive mode

built-in electric field forming the pulse of electric current in the input circuit of the operational amplifier.

The pulse rise time of the photodiode is determined by the processes of separation of charge carriers (electrons and holes) generated by ionizing radiation, by the built-in electric field strength and the p-n junction capacitance (Fig. 4.5). The time of charge carriers drift through the depletion region is proportional to its thickness and inversely proportional to the maximum velocity of the carriers in an electric field. For example, in germanium and silicon, the maximum drift velocity of charge carriers is around 5.10^6 cm/s, and the thickness of the p-n junction is usually less than 5 microns. Consequently, the time of charge carriers drift through the p-n junction is around 10^{-10} s. Detectors based on the p-n junction are usually fabricated by diffusion of p-type impurities into an n-type matrix or vice versa. The concentration of charge carriers in an impurity semiconductor is 10^{16}–10^{17} cm^{-3}. The radiation that occurs on the detector from the side of p-type semiconductor, allows one to create a larger number of electron-hole pairs. The thickness of the p-layer is selected so that the charge carriers have time to reach the p-n junction without recombination.

Besides segmented detectors, there are coaxial ring detectors, which allow selection of scattered electrons by the angle of scattering. These detectors, like the onlens detector described above, allows one to get different types of contrast. However, despite of some advantages, these detectors are not widely used.

Everhart-Thornley Detector

When recording true secondary electrons, a number of problems arise, one of which is that their energies are very small and lie in the range 0–50 eV (Fig. 4.2). The second problem is associated with the small spot diameter of the electron probe, which is about 1 nm. With such a spot, the current of secondary electrons does not exceed 0.01 nA. To register such low currents, Thomas Everhart, and Richard Thornley in 1960 proposed the design of detector, shown in Fig. 4.6. To detect a weak current, the scintillation detector is equipped with a number of additional devices. Since the electron energy must exceed 10 keV in order to generate a flash in the scintillator,

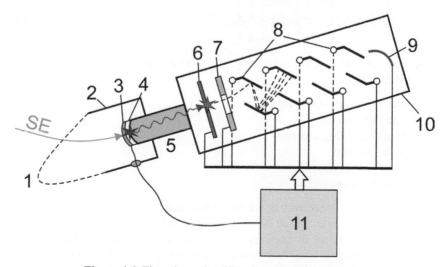

Figure 4.6. The schematic of Everhart-Thornley Detector

a thin translucent metal layer is deposited on its outer surface, on which a positive voltage of 12 kV is applied to collect and accelerate secondary electrons from the low-energy part of the spectrum. To exclude the influence of this field on the primary electrons of the probe, the scintillator is placed inside the Faraday cylinder, to which a voltage from −50 to +250 V is applied. Here a small positive potential is used to collect low-energy electrons, which, once inside the cylinder, are accelerated by an additional potential on the surface of the scintillator. Negative voltage at the collector is necessary in order to completely block the detector inlet for the low-energy part of the secondary electrons, thereby allowing the contrast to be observed only in backscattered electrons.

4.1.5   Types of contrasts in scanning electron microscopy

The image of the object viewed through an optical microscope does not differ much from that of the object observed by the naked eye. Therefore, the image of an object in an optical microscope does not need any interpretation. In SEM, the image is formed in a fundamentally different way, so the image observed on the monitor requires additional interpretation. To obtain an image in scanning electron microscopy, the same principle is used that was first implemented by one of the creators of modern television, Vladimir Zvorykin, in 1942. Vladimir Zvorykin created the idea of a scanning electron microscope, in which the role of the incident light was played by an electron beam moving along the sample from point to point, and the role of the reflected rays was played by secondary electrons. The image of the surface on the monitor was formed from luminous dots, the brightness of which was determined by the number of secondary electrons, their energy and the angle of scattering.

Thus, in electron microscopy, image quality, i.e., the distinguishability of individual surface details is determined by its contrast. A measure of the contrast of the image is the contrast ratio C, which is defined as follows. If at point A_1 the detected signal has value S_1, and at neighboring point A_2, respectively, S_2, then

$$C = 2 \frac{S_1 - S_2}{S_1 + S_2}. \tag{4.4}$$

In scanning electron microscopy, the following types of contrasts are distinguished:

- compositional or Z-contrast;
- topographic contrast;
- electron channeling contrast;
- magnetic contrast;
- potential (volt) contrast;
- orientational contrast.

For the correct interpretation of the image displayed at the SEM monitor, it is necessary to understand what kind of contrast determines one or another of its features. The main types of contrasts that are used to characterize the functional layers of solid-state lithium-ion batteries are compositional or Z-contrast and topographic contrast.

Compositional contrast

As noted above, the total current of backscattered electrons can reach 30% of the current of the electron probe. The backscattered electrons carry valuable information about the average atomic number at a given point in the object under study. The coefficient of emission of backscattered electrons η, determined by the differential scattering cross-section (Equation 4.3), depends on the atomic number and expands with the growth of the latter. If individual regions of the sample differ in their elemental composition, then the scattering coefficient of primary electrons of these regions will also be different. On the SEM monitor screen, areas containing elements with a high atomic number will appear brighter than areas containing lighter elements. This type of contrast in electron microscopy is called compositional or Z-contrast. Thus, it is possible to visualize the heterogeneities of the sample due to different elemental composition. This type of contrast is especially effective when examining transverse cleavage of SSLIBs, on which the functional layers of the battery are clearly visible. This type of contrast can be obtained using any kind of an electron detector.

Topographic contrast

The most effectively SEM methods are used to study surface topography. This type of contrast is due to the influence of the surface topography on two scattering parameters: the direction of motion of the backscattered electrons and their numbers. The number of electrons scattered in a certain direction is determined by the secondary emission coefficient η and depends on the orientation of the element on the scattering surface with respect to the electron probe. The secondary emission coefficient grows with an increasing angle of incidence of the ion probe — the angle between the ion probe and the local normal to the surface element of the sample. Therefore, if the detector could collect all the scattered electrons, then surface elements located at large angles to the electron probe would look brighter in the SEM image.

Since not all electrons reach the detector, the second parameter, the scattering direction, plays a decisive role in the formation of the topographic contrast. The largest number of reflected electrons lies in the plane formed by the direction of the electron probe and normal to the surface element. In this case, the angle of incidence is approximately equal to the angle of reflection. Thus, those faces for which the direction of the scattered electrons coincides with the direction to the detector will look the brightest. Of these, those located at the maximum angle to the electron probe will be brighter. This so-called orientation effect plays a more important role in the formation of topographic contrast than the secondary emission coefficient. In other words, the surface looks in more relief to a greater extent due to the orientational effect. It is important to understand that when one considers the orientation effect, it means that the orientation of the elements of the surface of the sample is relative to the detector. When one reflects on the role of the secondary electron emission coefficient, one means their orientation with respect to the electron probe.

The reliefs obtained on SEM images with a predominantly topographic contrast are similar to the images formed by an optical microscope, and the interpretation of the reliefs is largely similar to the interpretation of optical images. However, the quality of images obtained using SEM is always higher than in optical photographs, since the depth of the field in SEM is much higher than in optics. So at a magnification of ×500, the focus depth can reach ~ 0.5 mm, which is hundreds of times larger than in an optical microscope. This parameter is a very important distinguishing feature of a scanning microscope.

To obtain a topographic contrast at the SEM image, all the above types of detectors can be used, however, the best result is obtained using the inlens and four-segment semiconductor detectors. The latter can be used both to obtain Z-contrast, then the signals of all four segments A, B, C, D should be summed up, and to obtain the topographic contrast. In the latter case, the difference of any pairs of segments are summed, for example, $(A - B) + (C - D)$. When studying a relief surface, in one measurement event the maximum signal is received by one, at most two adjacent segments, to which the backscattered electron flow is directed. The signals of all segments contain components that in this case can be considered as a background, for example, signals of elastically and quasi elastically scattered electrons, which depend weakly on the topography of the surface. Subtracting the signals according to the above scheme eliminates this background and improves the signal-to-noise ratio in the mode of obtaining topographic contrast.

An Everhart-Thorneley detector can also be applied to produce a topographic contrast with backscattered electrons. To do this, a small negative voltage is applied to the Faraday cylinder, which cuts off real secondary electrons. Since the detector is located on one side of the sample, the solid angle at which the detector collects electrons will be very small, because only electrons scattered directly in the direction of the detector will reach it. Accordingly, only those areas that predominantly scatter electrons into the solid angle at which the detector aperture is 'visible' from electron beam spot will be bright in the image. Areas tilted to the electron probe at other angles will appear dark on the monitor screen. As a result, the contrast of the image will be very sharp, and the image itself will consist of dark and light spots and contain few areas with shades of gray.

Topographic contrast with secondary electrons

In the case of inelastic scattering of primary electrons on the outer shells of sample atoms, they are able to knock out secondary electrons from them. Some of them get kinetic energy sufficient to experience a series of inelastic collisions, cross the surface of the sample and exit into a vacuum. The electron of the primary beam has enough energy to knock out several such electrons. They are called true secondary electrons. In the N(E) dependence (Fig. 4.3), secondary electrons correspond to a maximum in the region from 0 eV to 50 eV. The criterion for classifying electrons as secondary is exclusively their kinetic energy – that is, all electrons emitted from a sample with an energy of less than 50 eV, despite the rather arbitrary choice of this threshold, are considered to be secondary.

As the energy of the secondary electrons is small, their output is possible only from the surface layers of the material. The maximum yield depth for metals is about 10 Å, and for dielectrics, about 100 Å. Usually most of the secondary electrons are emitted from the region lying at half the maximum electron exit depth. Due to the small kinetic energy, these electrons are easily deflected by a small potential difference. This allows in collecting the maximum possible number of electrons, which significantly increases the efficiency of the detectors and allows to get high-quality images with a good signal-to-noise ratio and a resolution better than 1 nm.

It should be noted that no systematic dependence of the magnitude of the secondary emission on the atomic number of the sample is observed. That is why the average value of the emission coefficient of true secondary electrons grows only slightly with increasing average value of the atomic number. Consequently, in secondary electrons, the contrast determined by the atomic composition of the target will not manifest itself. At the same time, the number of secondary electrons depends on the angle of incidence of the electron beam on the sample surface, that is, on the topography. Therefore, the image in secondary electrons is used to improve the topographic contrast, obtained using the Everhart-Thornley detector.

Since a positive voltage is applied to the Faraday cylinder to collect secondary electrons, along with the true secondary electrons, backscattered electrons also enter the detector. As noted above, the image in backscattered electrons obtained using the Everhart-Thornley detector appear very sharp and practically do not contain half tones. Such an image is very difficult to interpret using light-optical analogs, as part of the relief information disappears along with the half tones. To soften the contrast and increase the information content of the image allows secondary electrons. Since these electrons are more sensitive to a small potential at the detector, they are collected from the greater part of the probed surface and, as it were, backlighting the image. The image will look as if, together with a bright light source directed at a large angle to the normal to the plane of the sample, a scattered light source, for example, a flash lamp directed at the ceiling, was also used for illumination.

4.1.6 Practical use of SEM to study the functional layers of SSLIBs

The types of contrast listed above provide wide opportunities for the diagnosis of materials used in the manufacture of thin-film electrodes. When elaborating technology for PVD of functional layers of a solid-state lithium-ion battery, SEM is used for the following types of diagnostics:

- control of the morphology and continuity of the functional layers of the battery;
- visual control of the porosity of the electrodes;
- measurement of the thickness of functional layers;
- visual inspection of the of the functional layers' adhesion.

The study of the morphology of the SSLIB functional layers

When developing SSLIB manufacturing technology, first the deposition mode for each functional layer of the battery is worked out. At this stage electron microscopy allows to determine the quality of the adhesion of the functional layer to the substrate or other layers. In Fig. 4.7 the edge of the multilayer structure of $Si/SiO_2/LiPON/Li/Cu$ is shown. As can be seen in the figure in the process of chipping a copper film, acting as a current collector, is exfoliated. This is because the adhesion of lithium to LiPON and copper is greater than the cohesion of lithium itself.

The cleavage image in Fig. 4.7 was obtained at SEM Supra 40 (Carl Zeiss Group) using an inlens detector. The decisive role in the formation of a 'three-dimensional'

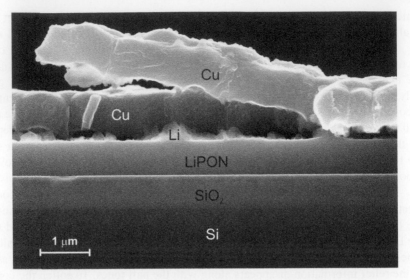

Figure 4.7. The cleavage of multilayer structure of Si/SiO₂/LiPON/Li/Cu

image in this figure belongs to topographic contrast. The selected angle conveys the volume of functional layers due to orientation effects, as well as due to the high electron emission coefficient of surface elements having a small radius of the curvature. In this case, the image looks as if the sample was uniformly illuminated from all sides, just like in optical dark-field microscopy. Due to this effect, a step formed by the SiO₂/LiPON and LiPON/Li/Cu layers is clearly visible. Additional brightness to this band is imparted by 'backlighting' by electrons reflected from the LiPON film and incident on the step at sliding angles.

It is interesting to note that in Fig. 4.7 there is no unambiguous dependence of the image brightness on the average atomic number, although the inlens detector provides Z-contrast. The materials present in the image, copper has the largest atomic mass, which, accordingly, looks like the brightest object. At the same time, the silicon cleavage appears darker compared to the film of silicon oxide and LiPON, although their average atomic mass is less than that of silicon. This is due to the fact that silicon dioxide and LiPON are dielectrics and accumulate a charge when scanning an electron probe.

Figure 4.7 shows the dark areas between the copper and LiPON layers, which are difficult to interpret based on data obtained by inlens detector. More useful in this case is a four-segment detector that provides Z-contrast. The corresponding image is shown in Fig. 4.8, where the brightness of the functional layers uniquely corresponds to the average atomic number of the layer. Here, the lightest layer is copper, and the darkest is lithium. Thus, what looks like a cavity in Fig. 4.7 actually turns out to be a lithium layer. Figure 4.9 shows this even more clearly, where on the top image (a) the cleavage is shown with topographic contrast, and on the bottom image (b) the same cleavage is shown with Z-contrast. In Figure 4.9a it is clearly seen that under the copper layer there is a certain substance, and at 4.9b it can be seen that this substance has an atomic number lower than that of LiPON, i.e., is lithium.

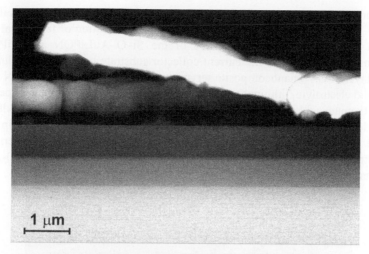

Figure 4.8. Image of a cleaved multilayer structure of Si/SiO$_2$/LIPON/Li/Cu, shown in Fig. 4.7. The image obtained with a four-segment semiconductor detector transmits Z-contrast more correctly

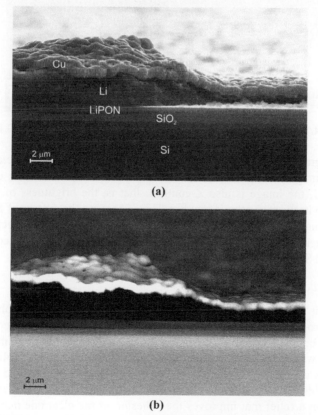

(a)

(b)

Figure 4.9. Chipped multilayer structure Si/SiO$_2$/Li/Cu/Ti: a - image obtained using the inlens detector; b - Z-contrast image obtained using a four-segment (onlens) detector

Figure 4.10 shows the 'cleavage' of the Ti/Si-O-Al/LiPON/Li/Cu multilayer structure. This structure was used for experimental verification of functional layers adhesion and measuring the resistance of the Si–O–Al/LiPON and LiPON/Li transitions. Here the role of the current collector substrate belongs to titanium foil, the five-layer Si-O-Al nanocomposite serves as the positive electrode, LiPON is used as the solid electrolyte, lithium serves as the negative electrode and the copper layer with the titanium sublayer serves as the second current collector. Since a flexible foil with a thickness of 10 μm was used as a substrate, it was not possible to chip such a structure. Figure 4.10 depicts the tear region of the foil, with exfoliated layers of the battery structure. It is interesting to note that the adhesion of LiPON to Si-O-Al was higher than that of all other materials of the test structure.

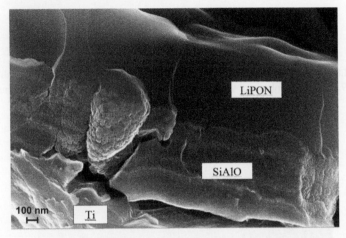

Figure 4.10. The tear region of multilayer structure Ti/Si-O-Al/LIPON/Li/Cu

The topographic contrast of the image obtained using the inlens detector, as in the previous images, completely conveys the depth and volume of the object. But unlike Fig. 4.7, the image is also Z-contrast, that is, the brightness of each layer corresponds to its average atomic number. This is because the system of Fig. 4.10 is grounded through a titanium collector, and all overlying layers have electronic and/ or ionic conductivity. This is enough for runoff of the electron charge accumulating on LiPON and Si-O-Al films.

Multilayer structures

In Chapter 3 it was noted that the Si-O-Al nanocomposite allows to obtain stable negative electrodes with a capacity of up to 2000 mA·h/g. Such a nanocomposite is formed by vacuum sputtering of silicon and aluminum targets from two magnetrons, the power of which changes into antiphase with simultaneous controlled oxygen supply. In this case a layered structure, shown in Fig. 4.11 is obtained. As a rule, the electrode films are deposited on a titanium foil (Fig. 4.11b), pretreated with an etchant to form a relief that improves the adhesion of the electrode material to the foil. Simultaneously with the electrodes on the foil, the same structure is formed on a silicon substrate. It is called 'witness' or test sample (Fig. 4.11a), necessary for control of some physical parameters of the film, such as conductivity, thickness, etc.

To study the morphology of layered negative electrodes, a Supra 40 SEM with an inlens detector was used. Though the atomic numbers of the main elements of the film differ by unit and the maximum difference in the concentration of silicon and aluminum does not exceed 50%, in Fig. 4.11 the individual layers forming the film are clearly distinguished. In the direction normal to the surface, the material has a low electronic conductivity since it contains layers with a high concentration of silicon dioxide and oxide. When interacting with an electron probe, a charge can accumulate in a region located at a considerable distance from the conductive substrate. Therefore, the lower region in Fig. 4.11a looks darker because it is located closer to a conductive silicon wafer located on a grounded table. The electrode layers adjacent directly to the substrate are also electrically neutral and therefore look dark. The upper edge of the film looks lighter since it is further from the grounded substrate and accumulates the charge. The same effect is observed for a film on a titanium foil (Fig. 4.11b) but is not so pronounced, because images obtained have higher brightness.

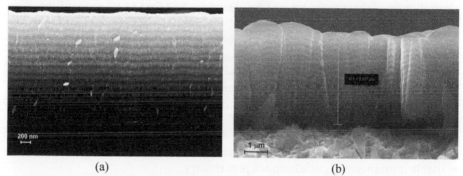

(a)   (b)

Figure 4.11. a - Layered negative electrode Si-O-Al on a silicon wafer. The film is charged unevenly, the upper layers are 'glowing' more. b - Layered negative electrode Si-O-Al about 4 μm thick on titanium foil. The light tones of the top of the film are also due to the accumulation of charge

Figure 4.12 shows the electrodes from the same series as in Fig. 4.11, but after several hundreds of charge-discharge cycles in the LP-71 electrolyte. First the film thickness, which was originally about 4 microns, is notable. As a result of cycling, the film swells to a thickness of ~25 μm (Fig. 4.12a) and ~16 μm (Fig. 4.12b) while the layered structure disappears. At the same time, the so-called Solid Electrolyte Interface (SEI) is formed on the film surface, consisting of lithium compounds with the products of electrolyte degradation.

Visual porosity control

One of the most important parameters of an electrode film is its porosity. Porosity is understood as the fraction of the pore volume in the total volume of the porous body. Porosity is a dimensionless quantity, varying from 0 to 100%. A material without pores corresponds to zero porosity, while a porosity of 100% is unattainable. Pores

(a)   (b)

Figure 4.12. Layered negative Si-O-Al electrodes on a titanium foil after cycling.
SEI layer is visible on film surface

can be either end-to-end or dead-end pore. For batteries with liquid electrolyte, the electrodes should have a through-pore structure to ensure uniform distribution of electrolyte throughout the volume of the electrode. And for SSLIB a porous structure is necessary for the development of the electrode/electrolyte interface and the mechanical unloading of the film. In this case, the type of pores is not so important.

The most obvious way to measure porosity is to measure the density of the sample $\rho_v = m/V$, where m is the mass of the sample with pores, and V is the volume of the sample with pores, and the true density of the sample material ρ_t. Then the porosity P can be calculated as $P = (1 - \rho_v/\rho_t)100\%$. Unfortunately for microporous thin-film electrodes with a thickness of 2-3 microns, this measuring method is practically inapplicable. In addition to it, there are about 60 other methods of measuring microporosity. The most common method implemented in commercially available instruments is gas adsorption porosimetry.

High porosity of the electrode films is achieved by special processing of the substrate surface and selection of the deposition rate determined by the power of the magnetron. At a low deposition rate and a smooth surface of the substrate, dense, uniform, polycrystalline films are formed. A correctly deposited electrode film should have a columnar structure and at the same time be amorphous, or at least X-ray amorphous. Such films are obtained at a sufficiently high rate of magnetron sputtering, which is controlled by the magnetron power and depends on the configuration of the magnetron-target-substrate system. As an example, Fig. 4.13 shows how the columnar structure of a vanadium oxide film looks immediately after its deposition (Fig. 4.13a). Figure 4.13b shows the same film after annealing in an oxygen atmosphere. Here as in the previous case, an inlens detector was used to obtain topographic contrast.

Films based on a Si-O-Al nanocomposite have substantially smaller pores (Fig. 4.14) the dimensions of which, nevertheless, allow the liquid electrolyte to freely penetrate deep into the film. When using a solid electrolyte, the porosity of the film does not significantly contribute to the development of the interface, since the electrolyte film does not penetrate pores whose diameter makes several nanometers. For this, pores with a diameter of about a hundred nanometers are needed. However,

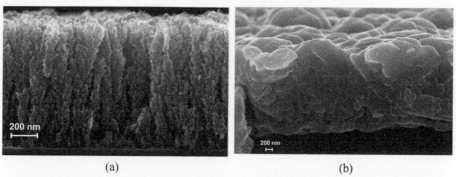

(a)   (b)

Figure 4.13. Cleavage of film: a - immediately after the vacuum magnetron deposition; b - after annealing in the atmosphere of rarefied oxygen. Image obtained with inlens detector

high porosity helps to increase adhesion, reduce mechanical stresses appearing from the introduction of lithium, and increasing the life of the film. The porosity can be increased by chemical etching [3] of the film in a solution based on hydrofluoric acid. A part of the Si-O-Al film volume is occupied by regions with a high content of silicon dioxide, which are washed out of the film in a hydrofluoric acid solution.

As already noted, the standard method for measuring porosity is gas adsorption porosimetry. However, the procedure for measuring porosity by this method is very long and laborious. At the same time, when working out the film deposition regimes, only a qualitative characterization of porosity is required, and exact porosity values are not necessary. Exact values are required when testing the electrodes to calculate the specific capacity of the electrode material, expressed in units of $A \cdot h/g$. Therefore, it is more convenient to use the SEM method to control porosity at the stage of technology development. Figure 4.14 shows the surface of a Si-O-Al film immediately after its deposition (a) and after treatment with a solution based on hydrofluoric acid (b).

The image was obtained using an inlens detector, which provides a topographic contrast sufficient for visual inspection of porosity. As can be seen in Fig. 4.14 a, the initial surface of the film consists of tubercles that are formed due to the artificially created roughness of the titanium foil. On the cleavage of the film (Fig. 4.11b) it can be seen that the film has a dense columnar structure. After etching (Fig. 4.14b) pores several nanometers in size are formed on the film surface.

Determination of linear dimensions of functional layers

SEM allows not only studying the morphology of films but is also an indispensable tool for measuring the thickness of layers, determining the size of pores and crystallites, as well as other structural and topological parameters of the SSLIB functional layers. In most cases, this information is necessary in order to correctly select the time of deposition of the layer by PVD methods. Adjusting the layer thickness by the power of the source – magnetron, thermal evaporator, etc., is possible, but undesirable, since in this case the particle deposition rate changes and, accordingly the film morphology changes as well.

(a)   (b)

Figure 4.14. The surface of the Si-O-Al film: a - immediately after deposition; b - after half an hour etching in a solution of hydrofluoric acid [6]

The software of a modern electron microscope provides ample opportunities for simple measurements of the studied objects, such as length, diameter, height, etc., as well as for more complex measurements – the perimeter and area of objects. Measurement can be performed both in manual and in automatic mode. The obtained information is displayed directly on the monitor on individual objects, as shown in Fig. 4.15, or on all objects identified in the field of view of the microscope. In this case, the dimensions are displayed in units corresponding to the magnification of the microscope: millimeters, micrometers or nanometers.

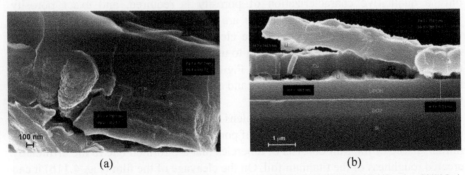

(a)   (b)

Figure 4.15. Cleavages of multilayered structures: a - Ti/Si-O-Al/LiPON/Li/Cu, b - Si/SiO$_2$/ LiPON/Li/Cu

For the correct display of sizes and ensuring the uniformity of measurements, calibration of the microscope on all used zoom ranges are required. For transmission electron microscopes, the magnification of which varies discretely, calibration is carried out separately for each magnification mode. For a scanning electron microscope, the magnification of which can change continuously, such a task is impracticable. Therefore, calibration is performed only at small magnifications with an error of ~ 1% and it is further believed that the magnification is proportional to the decrease in the size of the raster, which is determined by the amplitude of the scan voltage. In this sense, SEM is not a measuring tool, which does not exclude the possibility of making accurate measurements. For example, using working standards designed to calibrate measuring instruments.

Theoretically, the working standards for SEM verification can be provided by national institutes of standards. Since the SEM is not a measuring device and is not subject to mandatory metrological verification, working standards can be purchased from several markets of microscopy supplies. But usually, working standards are supplied by the SEM manufacturing companies along with the instrument. As a rule, these working measures are the property of the company and cannot be transferred to other users of measuring or analytical equipment. Alternatively, one can use the Standard Reference Material (SRM) 484 or NIST SRM 8091, available from the National Institute of Standards and Technology (USA, Boulder).

Figures 4.16 through 4.19 show examples of working standards supplied by the FEI company for SEM. Working standards are used to obtain a high-contrast image, which allows accurate setting markers when calibrating the microscope. For this, polished metal plates can be used with strokes applied with a diamond tool (Fig. 4.17), or with marks applied by lithographic method.

Figure 4.16. Container with working standards for SEM (FEI)

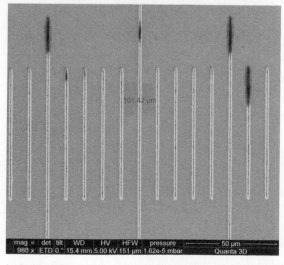

Figure 4.17. The FEI working standard in the form of a polished metal plate
with applied strokes on it

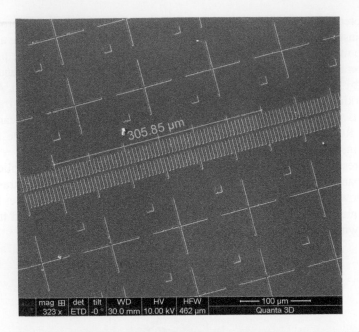

Figure 4.18. The FEI working standard with a photolithographically formed a dashed scale and two-dimensional topographic elements for calibrating the stigmator and the position of the aperture of the ion column

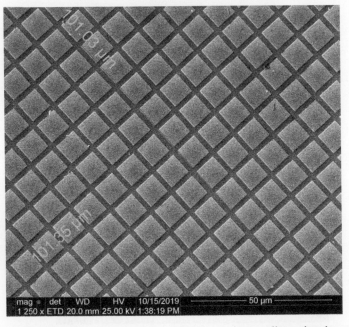

Figure 4.19. The FEI working standard in the form of a two-dimensional topographic relief with a period of 10 μm, made by photolithography

The calibration of the microscope is possible only after the specimen stage is set strictly perpendicular to the SEM optical axis. Otherwise, in view of the large depth of the field, all these strokes will be clearly visible, but the distances between them (division value) will be different. Moreover, the strokes themselves, which should look like a sequence of parallel stripes, will have a vanishing point.

The one-dimensional scale shown in Fig. 4.17 can be used to calibrate the magnification, but has the disadvantage that only one direction of the frame can be calibrated at a time. A rotation of the specimen stage by 90 degrees allows to calibrate the second direction, however, it is more convenient to carry out calibration in the presence of calibration grids (Figs. 4.18, 4.19). The grid allows to calibrate the orthogonality of the raster in one step. It is noteworthy that in scanning electron microscopes only relatively small magnifications are calibrated, for example, 1000 times. At higher magnifications, the raster is smaller, and a proportionally smaller signal is applied to the scanning coils. Since in the region of small current values, electromagnetic lenses have a more linear characteristic, additional calibration is not required. The dispersion of linear sizes for objects of size 100 nm or less in modern microscopes reaches 1.5–2%.

4.2 X-ray Spectroscopy

X-ray microanalysis or X-ray spectroscopy is an analytical method for elemental analysis or chemical characterization of a substance by probing the surface of a sample with an electron beam. The method has a high lateral resolution and allows analysis in very small volumes of the substance, the transverse dimensions of which on the surface under study are of the order of microns. For this analysis, special devices, called electron beam X-ray microanalyzers are used. As a rule, X-ray microanalyzers are built into scanning electron microscopes as additional attachments, designed to detect the characteristic X-ray radiation generated by the interaction of the electron probe with the atoms of the sample and containing basic information about its chemical composition.

4.2.1 The nature of characteristic X-rays

The characteristic X-ray was discovered by the English physicist Charles Glover Barkla in 1909, who subsequently (in 1917) was awarded the Nobel Prize in Physics for this discovery. In 1913, the British physicist Henry Moseley experimentally obtained a relationship linking λ - the characteristic X-ray wavelength with atomic number Z

$$\frac{1}{\lambda} = R(Z-\sigma)^2 \left(\frac{1}{n^2} - \frac{1}{k^2} \right) \tag{4.5}$$

Here R is the Rydberg constant (R = 1.1×10^7 m^{-1}), n is the number of the final energy level to which the electron has passed, k is the number of the initial energy level from which the electron has passed, and σ is the screening constant. From a formal point of view, expression (4.5) is a generalization of the Balmer-Bohr formula for a hydrogen-like atom in the case of atoms with two or more electrons.

Moseley's law, which explains the uniqueness of the characteristic X-ray spectra, forms the theoretical basis for the quantitative elemental analysis by the method of X-ray spectroscopy.

The mechanism for characteristic X-ray generating, which was already mentioned earlier, is as follows. Primary electron can knock electrons out of the inner atomic shells, forming a vacant energy level, also known as a core hole. These vacant levels are filled with electrons from one of the overlying levels with the release of energy in the form of an X-ray quantum, the frequency of which, according to Bohr's second postulate, is determined by the Equation $h\nu = E_n - E_m$. If an electron is knocked out from the lowest energy level 1s, then electrons from overlying energy levels can take its place, making radiative transitions $2p_{3/2} \rightarrow 1s$ and $2p_{1/2} \rightarrow 1s$ or $3p_{3/2} \rightarrow 1s$ and $3p_{1/2} \rightarrow 1s$ and so on. In Moseley's formula (4.5), these transitions correspond to the values of the principal quantum number $n = 1$ for the lower level and $k = 2,3$ for the overlying levels. Radiative transitions $2p_{3/2} \rightarrow 1s$ and $2p_{1/2} \rightarrow 1s$ or $3p_{3/2} \rightarrow 1s$ and $3p_{1/2} \rightarrow 1s$ are prohibited, as a change in the orbital quantum number l for them does not satisfy the selection rule of $\Delta l = \pm 1$.

In the transitions described above, the so-called K-series is emitted, consist of spectral lines $K_{\alpha1}$, $K_{\alpha2}$, $K_{\beta1}$, $K_{\beta2}$. For the designation of spectral lines from the notations of the Swedish physicist Karl Manne Georg Siegbahn, the Nobel Prize winner in 1924, were used. Within these notations the indices 1, 2 characterize the fine structure of the lines. If the electron is knocked out from the level $n = 2$, then the vacancies at the levels 2s and $2p_{3/2}$ will be filled from the higher levels $3d_{5/2} \rightarrow 2p_{3/2}$, $3d_{3/2} \rightarrow 2p_{3/2}$, $3d_{3/2} \rightarrow 2p_{1/2}$, $3p_{3/2} \rightarrow 2s$, generating an L-series consisting of lines $L_{\alpha1}$, $L_{\alpha2}$, $L_{\beta1}$, $L_{\beta2}$.

The energy levels, the transition between which is followed by emitting a definite line of characteristic X-ray spectrum, are shown in Fig. 4.20. The name of the X-ray lines is formed as follows. Each transition (or line) is indicated by a capital Latin letter, corresponding to the shell to which the transition is made (K, L, M ...), and a lowercase Greek letter in the index, indicating how the shell from which the transition is made corresponds to the one where the transition takes place (α, β, γ ...). For example, K_α indicates the transition to the K-level from the next overlying level, i.e., from level L. Symbol K_β denotes the transition to the K-level from the level M following the L level.

Figure 4.20 shows that the system of notations remains correct only for transitions to the K-level from levels L and M. Initially to designate X-ray lines a 'Zigban system' was used, developed many years ago by Manne Zigban and his students. The use of the letters K and L to denote X-ray lines goes back to 1911, by an article of Charles Glover Barkla entitled 'Spectra of fluorescence X-rays' [4]. But by 1913, Henry Moseley had already differentiated two types of K- and L-lines for each element, denoting them as α and β [5]. Therefore, it became necessary to take these types into account, for example, by introducing the corresponding indices in the notation of lines K_α and K_β. In 1914, while performing his dissertation research, Ivar Malmer, a student of M. Zigban, discovered that the α and β are not single lines, but are doublets. In 1916 M. Zigban published this result in the journal Nature, using the notation $K_{\alpha1}$, $K_{\alpha2}$, $K_{\beta1}$, $K_{\beta2}$, which soon came to be known as the Zigban notation [6].

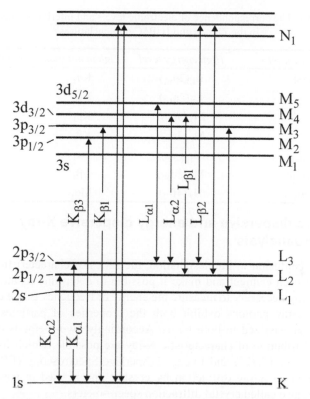

Figure 4.20. A schematic representation of energy levels and interlevel transitions, emitting characteristic X-ray quantum

It later became clear that the semi-empirical systematics of the so-called X-ray transitions introduced in this way do not take into account differences in the screening constant of shells and subtle effects in the atom. The proximity of the Mosley law and neglecting important effects in Bohr's theory led to the fact that the indices α, β, and γ do not reflect the correspondence of the spectral lines to the numbers of the electronic levels from which the transition occurs. Thus at present, there is a rather confusing notation system in which only the notation of the entire spectral series has an unambiguous relationship with the main quantum number, while the indices α, β, γ are arranged in accordance with historically accepted rules. The use of different approaches when designating the energy levels of K, L, M, etc., leads additional confusion. A number of authors consider them as electronic shells and designate them in accordance with their configuration 1s, 2s, $2p_{1/2}$, other authors use a set of quantum numbers [n, l, m].

Although Zigban notations are still widely used in spectroscopy, they are inconvenient and often misleading. For this reason, the International Union of Pure and Applied Chemistry (IUPAC) in 1991 proposed another nomenclature as an alternative to Zigban notations. Table 4.1 shows the correspondence of some of the most common transitions in the Zigban designations and in the IUPAC nomenclature.

Table 4.1. The correspondence of electronic levels and interlevel transitions in the Zigban and IUPAC nomenclatures

Low energy level	High energy level	Zigban notation	IUPAC notation
K (1s)	L_3 ($2p_{3/2}$)	$K\alpha_1$	K-L_3
	L_2 ($2p_{1/2}$)	$K\alpha_2$	K-L_2
	M_3 ($3p_{3/2}$)	$K\beta_1$	K-M_3
	M_2 ($3p_{1/2}$)	$K\beta_3$	K-M_2
L_3 ($2p_{3/2}$)	M_5 ($3d_{5/2}$)	$L\alpha_1$	L_3-M_5
L_2 ($2p_{1/2}$)	M_4 ($3d_{3/2}$)	$L\beta_1$	L_2-M_4
M_5 ($3d_{5/2}$)	N_7 ($4f_{7/2}$)	$M\alpha_1$	M_5-N_7

4.2.2   Wave dispersive and energy dispersive X-ray microanalysis

Despite some confusion in the designation, the series of characteristic X-ray lines are unique for each element and make it possible to determine its atomic number. For this, it is only necessary to measure the energy or frequency of the characteristic X-ray. Since X-ray photons exhibit both the properties of particles and waves, they can be characterized in both terms. Accordingly, two methods of measuring the spectral distribution of characteristic X-ray are possible, which form the basis of wave dispersive (WDS) and Energy Dispersive Spectroscopy (EDS). The first method is based on X-ray diffraction by crystals and is used in wave dispersive spectrometers, also called crystal diffraction spectrometers.

Wave dispersion spectroscopy

In crystal diffraction spectrometers, a crystal analyzer is used as a diffracting element. Characteristic X-ray from the sample falls on the crystal analyzer with a lattice constant d at an angle Θ. If the wavelength λ of any characteristic X-ray line satisfies the Bragg condition $2d\sin\Theta = n\lambda$ (the condition of the interference maximum, where $n = 1,2,3,\dots$), then the radiation intensity of this wavelength reflected in the direction Θ is amplified as a result of interference. The intensity of other characteristic X-ray lines for which this condition is not satisfied will be small or equal to zero. By moving the crystal analyzer and the detector (proportional counter) at different angles, it is possible to create conditions under which the condition of the interference maximum will be fulfilled and at the same time the 'reflected' X-rays will be focused on the window of the proportional counter.

In devices that use a focused electron beam (for example, in an electron microscope), X-ray radiation is generated in a small area on the surface of the sample. Therefore, the initially weak signal usually is maximized directly at the detection stage by using fully focusing X-ray spectrometers. An example of such a spectrometer is shown schematically in Fig. 4.21.

X-ray radiation hit a bent analyzing crystal, which together with a proportional counter can move along a circle of radius R. If the Bragg condition is fulfilled for some characteristic X-ray wavelength, then the intensity of this radiation scattered at

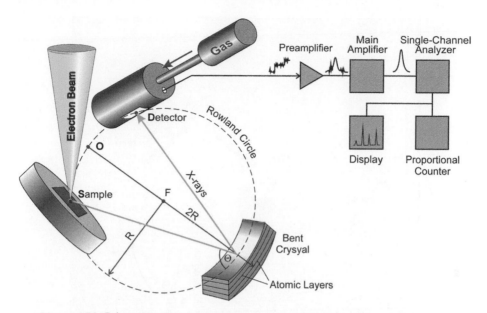

Figure 4.21. Schematic representation of a fully focusing X-ray spectrometer with a recording path

an angle Θ increases. After passing through the secondary collimator, this radiation enters the proportional detector and ionizes the gas in it (it can be a mixture of methane and argon or an argon-krypton mixture), resulting in Townsend discharge. This discharge forms an electric pulse whose amplitude is proportional to the energy of the X-ray quantum. After amplification and discrimination in amplitude, improving the signal-to-noise ratio, the signal itself is converted into a standard-size pulse by a Single-Channel Analyzer (SCA). The pulses generated in this way are counted by a pulses counter or displayed on the monitor as the X-ray spectrum i.e., dependence of the number of pulses on the angle of incidence. Each peak of the spectrum corresponds to a specific characteristic X-ray line.

In the spectrometer shown in Fig. 4.21, the so-called Johansson focusing is used when a point source of X-ray radiation, a crystal analyzer acting as a cylindrical mirror and a proportional counter lie on a circle of radius R, which is also called the focal circle or Rowland Circle (Fig. 4.22). If the radius of curvature of the mirror (in this case, a bent crystal) is 2R, then its focal length is R, and the focus itself is at point F (see auxiliary drawing in Fig. 4.22). If the point source of X-ray radiation S lies on the Rowland circle, and from this point S two rays come out — one parallel to the optical axis and the other passing through the focus F, then, reflected, they again intersect on the Rowland circle in point D. Consequently, all the rays, occurring on the crystal analyzer will be collected at this point. By placing an X-ray detector at this point, the optical power of the device can be increased significantly.

To make a bent crystal, it is initially ground so that it acquires a cylindrical shape with a radius of 2R, while the atomic layers remain flat. Then this is glued onto a cylindrical surface of radius R. As a result of bending, the atomic layers acting as a mirror acquire a curvature of radius 2R, so that the concave surface of the plate

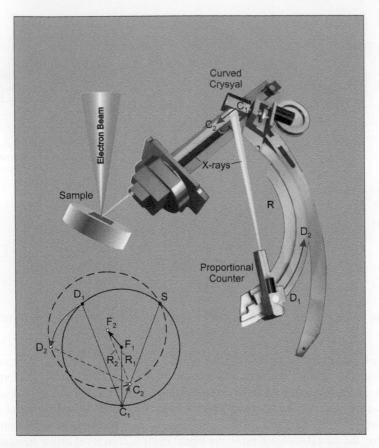

Figure 4.22. Focusing by Johansson. Red arrows indicate the displacement of the detector and the crystal. A drawing in the lower left corner explains the displacement of the Rowland circle when the detector and crystal move. Here C_1, D_1, F_1 are the initial positions, and C_2, D_2, F_2 are the final positions of the crystal, detector, and focus (center of the focusing circle)

becomes part of the Rowland circle. Since in the ratio $2d \sin \Theta = \lambda$ the range of Θ variation is limited by self-focusing conditions, and the range of λ variation is much wider than the range of $2d \sin \Theta$, crystals with different interplanar spacings d are necessary to fulfill the Bragg condition in the entire span of λ variation. Therefore, single crystals of quartz and lithium fluoride are used as crystal analyzers for short wavelengths (1–3 Å). For the analysis of X-rays in the long-wavelength range (10–50 Å), organic materials with a large interplanar spacing are used, for example, barium and lead stearates. In the ultra-long-wavelength X-ray range, where the characteristic X-ray radiation of Li and Be falls, diffraction gratings are used. As a result, the measurement process by WDS associated with the rotation and displacement of the detector and analyzing crystal, as well as the replacement of crystals, is very lengthy and it takes tens of minutes for X-ray spectral characterization. But the WDS remained the main instrument for X-ray spectral characterization until 1968, when SEM-based Energy Dispersive Spectrometers (EDS) appeared [1].

Energy dispersive spectroscopy

In energy dispersive spectrometers, unlike wavelength-dispersive spectrometers, there is no need for successive detecting of each spectral line. The EDS simultaneously processes all X-ray quanta falling into a semiconductor detector. The operation of EDS is based on the conversion of X-ray quanta to electrical pulses and counting the number of pulses with a given amplitude. Due to the unambiguous correspondence between the amplitude of the electric pulse and the energy of the detected X-ray quantum, it is possible to analyze the X-ray spectrum and determine the elemental composition of the specimen. The ratio of the number of pulses of a particular energy allows one to judge the relative concentration of the corresponding elements on the surface under study. By changing the energy of the electron probe, one can vary the penetration depth of the electron beam to a certain extent and obtain information on the distribution of elements in the surface layer.

In modern energy dispersive spectrometers, characteristic X-ray is registered by semiconductor detectors. The operation of such detectors is based on the phenomenon of the internal photoelectric effect and is in many ways similar to the detection of backscattered electrons in SEM (Fig. 4.3). At the same time, characteristic X-ray detectors are structurally different from BSE detectors, due to the greater penetration depth of X-rays. To obtain a good signal-to-noise ratio, the characteristic X-ray detector must have a larger volume and a larger thickness of the sensitive region, to ensure complete absorption of radiation in the sensitive layer. In this case, the material of the sensitive region must have a high resistance, otherwise the dark current fluctuations will be comparable in magnitude with the useful signal. Thus, the detector material must be either an intrinsic semiconductor or a compensated impurity semiconductor, or the Space Charge Region (SCR) can be used as a sensitive region.

Detectors in which SCR is used as the sensitive region are called surface-barriers. The surface-barrier detector (Fig. 4.23) is a diode structure formed by a thin p-silicon layer on the surface of n-silicon. The role of the contacts is performed by a gold film

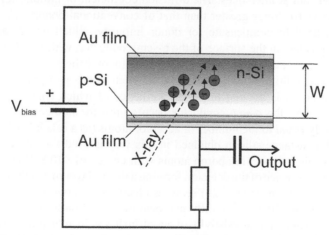

Figure 4.23. Surface-barrier X-ray detector. W is the width of the SCR or depletion region when reversely biased

deposited on the surface of the detector. The sensitive region is the SCR, which is formed as a result of diffusion of minority charge carriers: holes into n-silicon and electrons into p-silicon. Minority charge carriers recombination with the majority carriers results in the SCR formation in which there are no mobile charge carriers, and the space charge is created by uncompensated charges of impurities. These charges create a potential difference φ_v known as Volta potential, which prevents further diffusion of minority charge carriers.

The width of the space charge region W depends on the concentration of the donor N_d and acceptor N_a impurities, φ_v and the bias voltage V_b as

$$W = \sqrt{2\frac{\varepsilon\varepsilon_0}{e}(\varphi_v + V_b)\left(\frac{1}{N_d} + \frac{1}{N_a}\right)}. \tag{4.6}$$

For a surface-barrier detector the above conditions are always satisfied, therefore the expression for the SCR width can be presented in a more compact form 4.7

$$W \approx \sqrt{2\frac{\varepsilon\varepsilon_0}{eN_d}V_b} = \sqrt{2\varepsilon\varepsilon_0\mu\rho V_b} = 5,3\cdot10^{-5}\sqrt{\rho V_b}\ \text{cm}, \tag{4.7}$$

where μ is the electron mobility in silicon. It can be seen that with $\rho \sim 10^3$ Ohm×cm and $V_b \sim 10^3$ V the width of the SCR is $W \approx 5.3 \times 10^{-2}$ cm only. A further increase in bias voltage leads to an increase in detector noise. For this reason, surface-barrier detectors are not so widely used as X-ray detectors.

A particularly large volume of the sensitive region have a p–i–n-junction (more precisely, the n^+–p–p^+ junction where the '+' index means strong doping) detectors obtained by compensating the acceptor impurity with lithium ions. Lithium is used for compensation due to two main properties—it's extremely high mobility in tetravalent crystals and low ionization energy (0.033 eV in Si and 0.0043 eV in Ge). Due to their small size, lithium ions easily diffuse into silicon, being located not at the lattice sites, but at interstices. The diffusion coefficient of lithium, for example, in germanium is 10^7 times greater than that of conventional donors. This property of lithium is used to compensate for donor impurity, which occurs as follows. Lithium is deposited on the surface of the p-semiconductor, which is then heated to a temperature of about 400°C. When the heated lithium diffuses to a depth of about 0.1 mm, it forms a thin highly doped n^+ layer. Then to this p–n junction a voltage (plus to the n-layer) is applied under the influence of which controlled diffusion of lithium ions into the p-semiconductor occurs until the number of lithium ions becomes exactly equal to the number of acceptor atoms (as a rule it is boron).

The high-resistance silicon obtained in this way is used for fabrication p–i–n detectors. For this purpose, the boron atoms with energy of ~10 keV are implanted from the opposite surface of the detector, forming a thin p^+ layer of a high conductivity. Surface layers p^+ and n^+ (Fig. 4.24) serve as electrodes. Between them a sensitive region (up to 1 cm thick) of a fully compensated semiconductor is present, the resistivity of which equals a resistivity of an intrinsic semiconductor. The detector is placed in a sealed enclosure with a beryllium window through which X-ray radiation enters the sensitive layer of the detector. The completely absorbed X-ray photon of

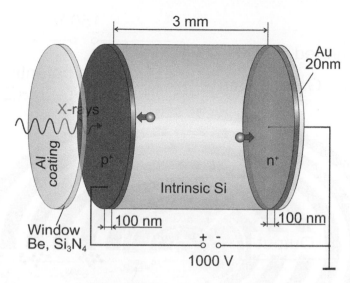

Figure 4.24. Detector with p-i-n-junction

E energy generates the electron-hole pairs, the average number of which is equal to $N = E/\varepsilon$, where $\varepsilon = 3.86$ eV is the average energy of an electron-hole pair formation at a temperature of liquid nitrogen. For example, an X-ray photon of copper with energy $E_{Si-K\alpha} = 8.05$ keV generates 2085 pairs, the charge of which is 3.337×10^{-16}C. This is a very small quantity invisible against the background of thermal fluctuations in the concentration of charge carriers (thermal noise) even in a fully compensated semiconductor. Therefore, to improve the signal-to-noise ratio, the detectors are used at liquid nitrogen temperatures.

In 1983, in a report by Emilio Gatti and Pavel Rehak 'The Semiconductor drift chamber – An application of a novel charge transport scheme' [7] the design of the Silicon Drift Detector (SDD), whose operation is based on the principle of lateral depletion was proposed. Figure 4.25 shows the design of an advanced version of the detector [8]. Its main advantage is a large volume of a sensitive or depletion region formed by n-type silicon with high resistivity. Depletion is attained due to the ohmic contact n^+, reverse biased with respect to the p^+ contacts, located on both sides of the sensor. The lower p^+ contact (Back Contact in Fig. 4.25) is deposited as a continuous layer, while the p^+ contacts on the opposite side are made in the form of concentric rings. The potentials on the rings are set by the built-in voltage divider, so the voltage from an external source is supplied only to the inner and outer rings. Concentric rings create in the depletion region a radial electric field, parallel to the surface. Electrons are generated by ionizing radiation and thermally excited electrons (the dark current), drift in the field to the n^+ contact of the substrate, which acts as a collecting anode. The configuration of the electric field is such that the holes drift in the direction of the annular p^+ contacts and are captured by the last ring.

Figure 4.26 shows the SDD equivalent circuit. According to the scheme, the electric current, appearing in a course of the charges separation by an electric field, decreases resistance of a reverse biased diode and a potential drop, occurring on the

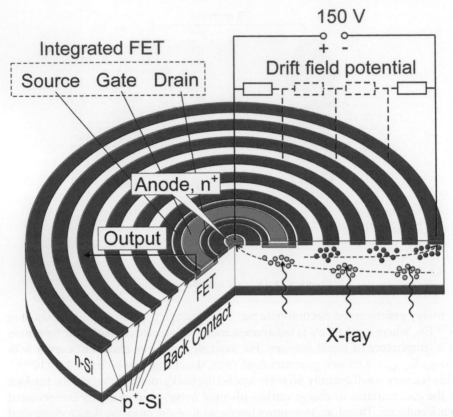

Figure 4.25. Design of a silicon drift detector based on the principle of lateral depletion [8]

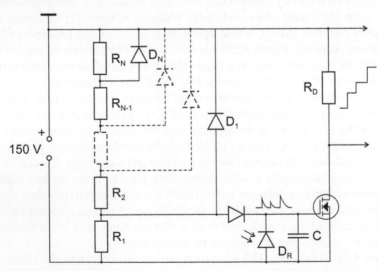

Figure 4.26. $R_1 - R_N$ is a voltage divider, $D_1 - D_N$ are diode structures formed by p+ rings and n+ anode (Fig. 4.24)

resistor charges the capacitor C. A stepwise increase in the voltage at the gate of the field-effect n-channel transistor leads to a stepwise increase in the voltage drop at the drain resistor. Next the signal is fed to the input of the preamplifier, as shown in Fig. 4.26. Since the charge on the gate can only increase until the transistor enters saturation mode, it is necessary to periodically reset the voltage on the capacitor C. For this purpose, a photodiode is used, which shortens capacitor C when irradiated with a LED. All electronic components shown in Fig. 4.26, integrate into SDD.

Artifact inherent with silicon detectors

In the process of X-ray photons being detected at the initial stage of ionization an electron can be knocked out from any shell, including the K-shell. In this case as already described earlier, the relaxation of an electron from the L-shell leads either to the emission of an $Si - K_\alpha$ X-ray photon or to the generation of an Auger electron. If the Auger electron is located more than 1 μm from the boundary, then it is most likely that it will be absorbed by the atoms of the detector with the formation of new electron-hole pairs. As a result, their number will not change and will remain equal to $N = E/\varepsilon$ where $\varepsilon = 3.86$ eV. If however, as a result of electron transition from the L-shell to the K-shell an X-ray photon is emitted, then two cases are possible. In the first case, the $Si - K_\alpha$ photon can knock out an electron from the L-shell of another Si atom. This electron will be replaced by an electron from the M-shell, etc., and as a result, the average number of electron-hole pairs will remain unchanged and equal $N = E/\varepsilon$. In the second case, the $Si - K_\alpha$ X-ray photon leaves the detector, because X-ray radiation has a higher penetrating power than an Auger electron of the same energy. If the $Si - K_\alpha$ photon is present near one or another detector surface, such an event is most likely. In this case, the number of electron-hole pairs generated by the X-ray photon is calculated by the same formula, but, taking into account the fact that part of the energy of the primary photon, equal to the energy of $Si - K_\alpha$ photon $E_{Si - K\alpha} = 1.74$ kcV, leaves the detector volume. In this case, the number of generated electron-hole pairs will be equal to $N_{esc} = (E - E_{Si-K\alpha})/\varepsilon$.

A decrease in the number of electron-hole pairs resulting from the loss of an X-ray quantum leads to the appearance in EDS spectrum of an artifact called 'escape peak'. It means that instead of a part of the pulses with amplitude $u \sim E/\varepsilon$, a number of pulses of $u_{cse} \sim (E - E_{si-k\alpha})/\varepsilon$ amplitude appear at the output of the detector. As a result, an additional line appears on the EDS spectrum, the energy of which is equal to the energy of the parent line minus 1.74 keV. Generally speaking, escape peaks appear from both $Si - K_\alpha$ and $Si - K_\beta$ photons, but the probability of their formation from $Si - K_\beta$ is approximately 2% of the probability of formation from $Si - K_\alpha$, therefore usually there is only one escape peak per one parent peak. The amplitude of the escape peaks with respect to the parent K-peak varies from 1.8% for phosphorus to 0.01% zinc. Silicon escape peaks cannot appear for photons whose energies are lower than the excitation energy of the silicon K-shell (1.838 keV) [1].

4.2.3 Principle of operation of SSD based EDS

The principle of the EDS operation on the example of SDD is illustrated in Fig. 4.27. The SDD chip bonded with a ceramic plate is mounted on a Peltier thermoelectric cooler and encapsulated in a sealed, evacuated case. The temperature of the SDD is controlled by a diode located directly on the detector and acting as a temperature sensor. X-ray photons enter the detector through a thin window of beryllium or silicon nitride, which isolates the detector case from the environment. The absorption of each individual X-ray photon leads to the formation of electron-hole pairs, spatially separated by a built-in electric field. In this case, an electric current appears in the circuit of the SDD anode, which charges a capacitor connected to the gate of the field-effect transistor located directly on the detector (insert in Fig. 4.27). A stepwise change in the potential at the gate of the field-effect transistor is converted by it into current pulses, and then the preliminary amplifier converts them into voltage pulses. The amplitude of each pulse is proportional to the value of the step at the input of the pre-amplifier, i.e., photon energy. Then these pulses are shaped and amplified by the main (linear) amplifier, which transmits them to a Computer X-ray Analyzer (CXA).

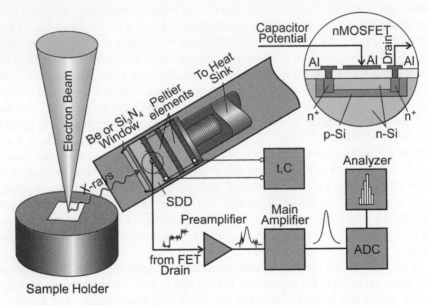

Figure 4.27. Schematic representation of energy-dispersive spectrometer and associated electronics. ADC is Analog to Digital Convertor. Depth of cooling by stage Peltier elements makes from –20°C to –80°C

The purpose of the analyzer is to sort the pulses and accumulate in the channels information on the number of pulses of the corresponding amplitude, i.e., counting how many pulses have an amplitude in the interval $V \div V + \Delta V$. For example, if the analyzer has 1024 channels, and the maximum amplitude of the pulse is 5.12 V, then the interval $0 - 5.12$ V is divided into 1024 channels and the discretization of the pulse amplitude is $\Delta V = 5.0$ mV.

The results of the detection of X-ray radiation are presented by a sample in the form of a table (Table 4.2). In the first row of the table the numbers of channels which are of interest are indicated and in which pulses of the corresponding amplitude should be counted. In the second line the interval of the amplitudes of the pulses that are counted in this channel are presented. In the third line the contents of these channels – the number of pulses accumulated over a certain time are indicated. In the fourth line, the energy range of X-ray photons corresponding to the amplitude interval are given. Assuming that as a result of a short-term turn on of the detector, a certain number of pulses were accumulated in the analyzer channels, and these pulses were distributed among the channels as indicated in the second and third rows of Table 4.2.

Table 4.2. An example of pulses distribution over channels

Channel number	760	770	780	790	800	810	820	830	840	850
The amplitude of the pulse, V	3.800 3.805	3.850 3.850	3.900 3905	3.950 3.955	4.000 4.005	4.050 4.055	4.100 4.105	4.150 4.155	4.200 4.205	4.250 4.255
Number of pulses	144	147	183	333	451	433	320	199	135	112
Photon energy, keV	7.600 7.610	7.700 7.710	7.800 7.810	7.900 7.910	8.000 8.010	8.100 8.110	8.200 8.210	8.300 8.310	8.400 8.410	8.500 8.510

The table data presented as a plot of the number of pulses versus photon energy, as shown in Fig. 4.28a. The figure shows that the experimental points form a certain spectral line, the origin of which is still unknown. Knowing the photon energy (in this case, 8.04 keV), one can find the sequence number of the corresponding element using the Moseley formula (4.5), provided, of course, the screening constant is known. In this case, if one assumes n = 1 and k = 2, the Mosley formula gives Z = 28.02 + σ which corresponds to Cu – K_α and σ = 0.98.

However, for a more precise attribution of the spectral line, it is enough to simply plot the entire spectrum or part of it, as shown in Fig. 4.28b. Then, next to the first peak, the second peak of 8.90 keV will appear, allowing one to assume with sufficient confidence that this is a doublet Cu – K_α and Cu – K_β. In this way, the required span of spectrum can be depicted by displaying the content of all 1024 channels on the analyzer monitor and identifying the peaks 'manually'. Modern energy dispersive spectrometers allow the identification of spectral lines in both 'manual' and automatic modes.

In conclusion, it should be noted that the resolution of the wave dispersive spectrometer is significantly higher than the energy dispersive one. So for example, the photon Si–K_α energy is 1.73 keV. After detecting X-ray photon by EDS, the half-width of the line will already be 0.1 – 0.15 keV. That is, the resolution of the EDS is 0.1–0.15 keV, while the half-width of the same line in the spectrum of the WDS does not exceed 0.01 keV. Such a narrowing of the energy window of the instrument can significantly reduce the background created by bremsstrahlung (brake radiation) and increase the sensitivity of the device. Figure 4.29 shows the peaks of silicon and aluminum that are components of the nanocomposite electrode material Si-O-Al, obtained by energy dispersive analyzer Apollo X. Ibidem, for comparison, it is shown how these lines would look on the spectrum of a wave dispersive spectrometer.

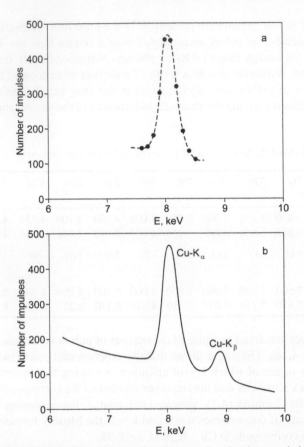

Figure 4.28. Graphical display of the results of processing the characteristic X-ray spectrum by a multi-channel computer X-ray amplitude analyzer: a - spectrum, plotted for ten points from Table 4.2; b - spectrum, plotted for 1024 points

However, energy dispersive microanalysis is interesting in that X-ray photons of all energies are detected simultaneously, not in turn, like in WDS. So for the EDS, the measurement process takes minutes in contrast with WDS, where it is lasts for tens of minutes. Thus the EDS, which allows getting the entire spectrum at the same time, but with a lower resolution, is more suitable for qualitative analysis. The WD analysis, although more laborious, enables higher resolution and is used for quantitative elemental analysis. Very often EDS is used for fast elemental scanning to find out what the material contains, and then it is already examined using WDS to obtain accurate data on its chemical composition.

4.2.4   The procedure of ED analysis and format for the presentation of measurement results

Automatic qualitative ED analysis

The automatic qualitative analysis system is an expert system that simulates the expert's behavior and reproduces the main stages of the 'manual' qualitative analysis

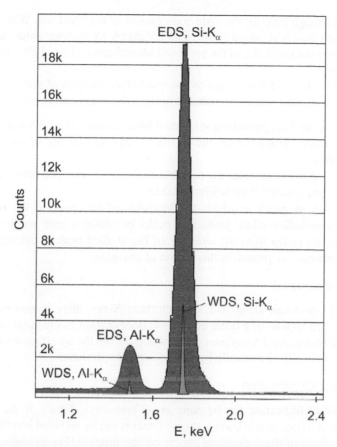

Figure 4.29. Illustration of the EDS (dark peaks) and WDS (light peaks) resolutions on the example of and lines of the Si-O-Al nanocomposite

carried out by the expert in recognition and classification of peaks. Like any expert system, it consists of two subsystems – an inference engine and a knowledge base. The inference engine is designed to apply logical rules to the knowledge base to obtain new information. The knowledge base contains information about the ED spectra of various elements and a set of certain rules that allows one to draw logical conclusions based on available facts.

There is a variety of automatic qualitative analysis systems from the simplest to complex. The simplest systems represent data obtained in a given energy range, practically without subjecting them to comprehensive analysis, intended to eliminate the risk of incorrect identification (assignments) of peaks. Almost all operations of 'manual' qualitative analysis in these systems are performed by the operator himself. Higher-level instruments use true expert systems with strict codification of qualitative analysis guides to avoid erroneous identifications. In such systems, the components of the spectrum are often given in decreasing order of reliability of the result. These systems allow peak identification in both manual and automatic mode. Moreover the degree of reliability of the qualitative identification of peaks will be high enough only for the main components of the spectrum. With respect to secondary and poorly resolved peaks, it is necessary to observe some caution and when assessing the reliability of the proposed identification, first of all, proceed with common sense.

With the variety of approaches to the qualitative analysis of energy dispersive X-ray spectra, its main stages are reduced to the following operations:

(1) Removing the background using spectral filters or background X-ray modeling.
(2) Peaks search, during which the positions and amplitudes of the peaks are determined.
(3) Peaks recognition, in which the possible elements to which certain peaks can be assigned are selected from reference tables.
(4) Subtraction of peaks at which the intensity of the peak of the recognized element, including minor peaks and peaks of artifacts, such as the escape Si peak, is subtracted from the intensity of the studied peak to determine if any other elements are present in this region of energies.

Background removal

In the X-ray spectrum, along with characteristic X-ray, there is also background radiation, which appears as a result of electron deceleration inside the sample. In the analysis, the background X-ray must be subtracted from the spectrum intensity. This is done automatically or manually by smoothing the spectrum.

Manual peak identification

Manual peak identification can be done in several ways. Firstly, if the elemental composition of a specimen is known, these elements can be included into the database and then markers of these elements appear on the monitor. For example, Fig. 4.30 shows the ED spectrum of a solid electrolyte LiPON with platinum electrodes for which markers of platinum are displayed to indicate Pt peaks possible positions. The operator can only make sure that there is indeed a peak at the marker position. If the spectrum contains unknown peaks, then clicking on them with the cursor, one can get a list of elements to which these peaks may belong, arranged according to the degree of relevance. From this list, the operator can select the element that, in his opinion, most corresponds to this peak, and add it to the list of identified elements. If it turns out that the alleged presence of any element in the sample contradicts the experimental data, then it can be excluded from the list.

Figure 4.31 shows an example of identification of the energetic spectrum of a thin-film negative electrode with a thickness of 3 μm. The active substance of the electrode is a Si-O-Al nanocomposite deposited by magnetron sputtering of Si and Al targets in a nitrogen-oxygen mixture onto a titanium foil. The energy of the electron beam in the analysis was 10 keV. The elemental composition of the nanocomposite is shown in Table 4.3. It also contains the so-called ZAF-corrections that take into account atomic numbers of elements and absorption coefficients in a particular sample.

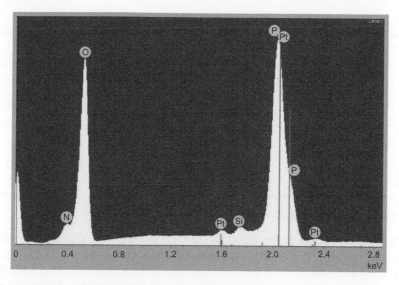

Figure 4.30. ED spectrum of LiPON solid electrolyte with platinum electrodes. Thin lines are platinum markers, where the marker height is proportional to the line intensity. The Si-K$_\alpha$ peak appears due to silicon detector atoms fluorescence

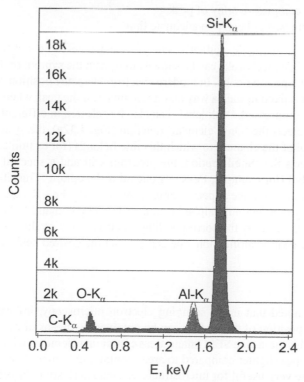

Figure 4.31. Energy spectrum of a Si-O-Al nanocomposite deposited on a titanium foil. The energy of the electron probe is 10 keV. Titanium lines not shown

Table 4.3. The elemental composition of the nanocomposite Si-O-Al

Element	Wt %	At %	K-Ratio	Z	A	F
C-K	3,3	6,83	0,0033	1,0628	0,0938	1,0003
Ti-L	13,91	7,22	0,0488	0,9019	0,3884	1,0009
O-K	18,59	28,89	0,031	1,0465	0,1595	1,0007
Al-K	7,82	7,21	0,0554	0,9775	0,7044	1,0281
Si-K	56,14	49,72	0,4044	1,0066	0,7149	1,001
Ti-K	0,23	0,12	0,0019	0,898	0,9198	1
Total	100	100				

Automatic spectrum identification

Automatic identification of the spectrum can be performed both during the spectrum acquisition and on completion of the spectrum acquisition. In the first case, it is nevertheless necessary to wait a reasonable time to achieve the necessary statistics, and then the identification of the peaks will be more accurate. Automatic peak identification is useful when:

- all peaks are well defined;
- there is no complex overlay of peaks;
- the spectrum was collected in adequate time.

The results of automatic qualitative analysis can be presented in different formats (Fig. 4.32). For instance, peaks can be shown along with the symbol of the element to which this or that peak belongs. In modern spectrometers, the position of the symbol in the spectrum is fixed in such a way that it remains near the peak when the spectrum is stretched or compressed. Peak labels can also be selected in different ways. If, for example, one selects the 'Only element' function (Fig. 4.32a), the system will show only the symbol of the element to which this line belongs (for example, Fe).

If one selects the 'Shell' option, the program will add a spectral series to the element's symbol (for example, Fe-K). If one adds the option 'Transition', then the full name of the spectral line appears, for example $Fe - K_\alpha$, $Fe - K_\beta$, etc. (Fig. 4.32b). If it is necessary to select only alpha lines, then by the command 'Alpha lines only' the program will sign only the corresponding spectrum lines, as shown in Fig. 4.32c. With automatic peak identification, one can also set the desired peak-to-background ratio.

Mapping

Earlier it was noted that using scanning electron microscopy, information on the elemental composition of the surface can be obtained by means of compositional or Z-contrast. In this mode, the SEM image areas containing elements with a high atomic number will appear lighter compared to areas containing lighter elements. Although such images are very useful for microstructures characterization, information on the elemental composition of the region is exceptionally a qualitative nature and is as vague as the image itself. However, if instead of scattered electrons a characteristic

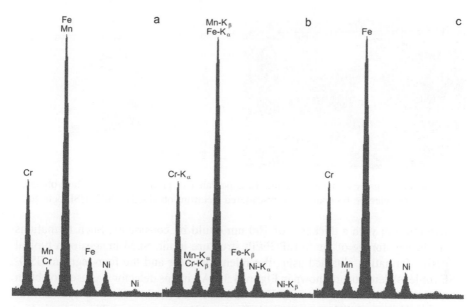

Figure 4.32. Examples of different formats for the presentation of an automatic qualitative analysis results

X-ray is used for imaging, then a fairly reliable picture of the surface distribution of an element, starting with boron ($Z = 5$), can be obtained. In X-ray microanalysis this method is known as 'X-ray mapping'.

The essence of the method is that wave dispersive and energy dispersive analyzers can be tuned to a specific chemical element with atomic number $Z \geq 5$. In the first case, one needs to tune to a specific wavelength λ, which is associated with atomic number Z by a Mosley formula. To do this, the analyzer should be adjusted to an angle $\Theta = \arcsin(\lambda/2d)$. In the second case, the analyzer needs to be tuned to a certain energy, by setting the required 'energy window' with a help of a discriminator. By tuning the analyzer to the desired element, one can scan the surface of the sample with an electronic probe. Whenever the electron beam enters the region where the concentration of the desired element exceeds the sensitivity of the device, a voltage pulse will be formed at the amplifier output.

By displaying on the monitor, the position of the electron probe on the surface of the sample and modulating the brightness at each point of the image with a signal from the amplifier output, one can get an image of this element distribution over the surface or an X-ray map. For example, one can put a white dot on the monitor screen, every time an impulse comes from the output of the amplifier, meaning there is an element of interest to one in this place of the sample. Despite the difficulties in obtaining high-quality X-ray images due to a weak X-ray signal, X-ray mapping quickly became one of the most favored methods for conducting X-ray spectral analysis. The reason for the popularity of the mapping method lies in its visibility and the convenience of visualizing the distribution of chemical elements.

As an example of the X-ray mapping application, maps of extrusive formation on the surface of a solid LiPON electrolyte with a thickness of 1 μm, covered by

Figure 4.33. Extrusive formations having the appearance of prominences: a - image obtained using inlens detector; b - image in backscattered electrons obtained by SSD-BSE detector

a platinum film with a thickness of 100 nm could be considered. Such formations occur during storage of the Si/LiPON/Pt structure in air. SEM images of a typical extrusive formation obtained using the inlens detector and the four-segment SSD-BSE (onlens) detector are shown in Figure 4.33. An inlens detector was used to obtain a topographic contrast, while SSD-BSE detector was used to obtain Z-contrast, for which the signals of all four segments A, B, C and D were summed. Therefore, in Fig. 4.33b, regions containing light atoms appear darker than the surface of the sample coated with a platinum film. Such light lithium compounds may be hydroxide, nitride and carbonate. To confirm this assumption, one needs to consider maps of extrusive formation.

Figure 4.34a shows of one of these prominences at zero tilt. Figures. 4.34b – 4.34g show the images of the selected area (white square) obtained in the following monochrome X-ray radiation: b in $C - K_{\alpha 1}$; c in $N - K_{\alpha 1-2}$; d in $O - K_{\alpha 1-2}$; e in $P - K_{\alpha 1}$; f in $Pt - M_{\alpha 1}$; g in $Si - K_{\alpha 1}$.

The considered example is interesting in that the method of X-ray spectral analysis does not allow directly to 'see' lithium, because the energy of its characteristic X-ray radiation is very small – less than the half-width of the line of the semiconductor detector. But one can see the distribution of the other elements that make up the sample and are contained in the atmosphere and based on their distribution to judge the distribution of lithium.

The similarities in the distribution of carbon (b) and oxygen (d) suggest that the bright region is lithium carbonate. It is also possible that oxygen in the composition of lithium hydroxide contributes to the bright region on map 'd'. It is formed where lithium emerges on the surface of a platinum film as a result of lithium interaction with carbon dioxide in the presence of water vapor. The pure platinum on maps 'b' and 'd' appears as a dark area in the upper left corner, that can be seen when comparing them with map 'f'. The relatively uniform distribution of nitrogen over the surface (c) is explained by both its rather high content in LiPON (8% and more) and the possible formation of lithium nitride on interaction with atmospheric nitrogen. The reasons for phosphorus (e) and platinum (f) presence are quite obvious. The presence of silicon (g) is explained by the small thickness of the LiPON film, which is permeable to the electron probe. More interesting is that, on the maps of these elements (e, f, g,), the entire right side is dark. This is due to the fact that lithium 'protuberance' more

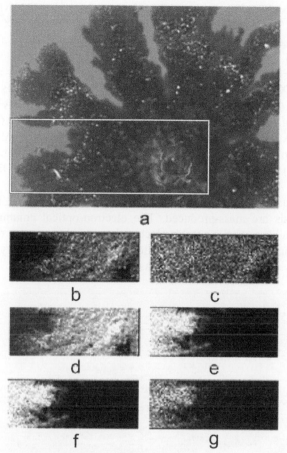

Figure 4.34. Maps of the elements of one of the spots on the surface of the solid electrolyte LiPON. The bottom inserts are images of the selected area obtained in X-ray monochromatic radiation: b in C - $K_{\alpha 1}$; c in N - $K_{\alpha 1-2}$; d in O - $K_{\alpha 1-2}$; e in P - $K_{\alpha 1}$; f in Pt - $M_{\alpha 1}$; g in Si - $K_{\alpha 1}$

than 10 μm height (Fig. 4.33) shadows the underlying areas from the electron probe. Thus, X-ray spectral analysis allows one to study the planar distribution of even those chemical elements that are "invisible" to X-ray spectrometers.

4.2.5 Industrial spectrometers for X-ray microanalysis

Structurally, X-ray spectrometers (EDS and WDS) are usually designed in the form of an attachment for scanning or transmission microscopes. At the same time, developers are trying to adapt their design to the widest possible range of microscopes. Recently Oxford Instruments produced X-ray spectrometers (INCAx-act and INCA Wave), which were intended for scanning, for example, Carl Zeiss (Fig. 4.35), and transmission microscopes. Now the company has developed the AZtec system, which is a hardware-software complex based on X-Max detectors and replaced the earlier version of the INCA system. For elemental analysis inside transmission electron microscopes, Oxford Instruments developed the high-performance silicon-drift detectors X-Max80T (with an ultra-thin polymer window) and X-Max80TLE (without a window). The Bruker company is producing the QUANTAX EDS and WDS systems for high resolution X-ray microanalysis on the basis of a scanning electron microscopes. In particular, TESCAN microscopes are equipped with these systems. SEMs by FEI company production are completed with EDS and WDS systems by the EDAX Company. For instance, energy-dispersive analysis of the Si-O-Al nanocomposite (Fig. 4.31) was performed at SEM Quanta 3D 200i with EDAX ED-spectrometer (Fig. 4.36). The ED-analysis of LiPON solid electrolyte (Fig. 4.30) was carried by INCAx-act EDS (Oxford Instruments) attachment to SEM Supra 40 (Carl Zeiss), shown in Fig. 4.35. SEM images of extrusion formations (Fig. 4.33) and their maps (Fig. 4.34) were also obtained at SEM Supra 40 equipped with INCAx-act EDS-console.

In addition, scanning electron microscopes especially designed for X-ray spectral microanalysis are mass-produced. The electron-optical column of such devices makes it possible to obtain a stable electron beam with an energy of up to 50 keV at the probe's currents of up to several microamps. They are usually equipped with several WDS systems - up to 5-6 pieces, as well as one EDS.

Figure 4.35. Microscope Supra 40 with an energy dispersive analyzer INCAx-act

4.3   X-ray structure analysis and X-ray phase analysis

X-rays are short electromagnetic waves with wavelength from $\sim 10^{-7}$ to $\sim 10^{-12}$ m ($\sim 10^3$ to $\sim 10^{-2}$ Å). On the scale of electromagnetic waves, this corresponds to the interval between ultraviolet radiation and gamma radiation. The energy of X-ray quanta lies in the range from ~ 10 eV to ~ 1 MeV. X-ray radiation is generated

Figure 4.36. SEM Quanta 3D 200i with an Apollo X energy dispersive analyzer

during the deceleration of electrons in the field of heavy particles and as a result of electronic transitions between atomic orbitals. The latter mechanism, responsible for the generation of characteristic X-ray, was discussed in detail earlier. The wavelength of characteristic X-ray Cu–K$_\alpha$ generated during the bombardment of copper by electrons with energy of ~ 30 keV is 1.5418 Å. This detail is of importance because copper is most commonly used as the anode material of vacuum diode also known as X-ray tube.

Substances that have a crystalline structure can be considered as a set of atomic planes, the distances between are comparable to the X-ray wavelength. These planes can reflect X-rays, which, depending on the angle of incidence, will amplify or weaken each other. This phenomenon, known as diffraction, as applied to X-rays, was discovered by Max von Laue and his co-workers Walter Friedrich and Paul Knipping in 1912. The theoretical rationale for this phenomenon was given by William Lawrence Bragg. It should be noted that some terminological confusion appears, when the phenomena of interference and diffraction are referred to by the same term as 'diffraction'.

Initially, diffraction was understood as the waves' ability to bend around the obstacles, the dimensions of which are comparable to the wavelength and propagate into the region of the geometrical shadow of the obstacle. In the case of X-rays, the reflection from the interatomic planes is of the same nature. The alternation of light and dark bands or rings (or more complicated figures in the case of mechanical waves) observed behind the obstacles was denoted by the term 'interference', implying the amplification or attenuation of the wave amplitude at a given point in space as a result of a superposition of waves with a certain phase difference. The alternation of light and dark bands or rings was called the interference pattern. Later, these terminological differences between 'interference' and 'diffraction' gradually disappeared and currently both phenomena are known by the term 'diffraction'. Returning to coherent X-ray scattering on crystalline structures, it should be noted that atoms are exactly the same obstacles for X-rays, therefore, when they are 'bent around', interference (or diffraction) patterns can also be observed. There are two approaches to studying the crystal structure based on different methods of obtaining interference pattern.

The Laue method is used to study the crystal structure of single crystals. The sample is irradiated with an X-ray beam with a continuous emission spectrum. In this case, the mutual orientation of the beam and crystal does not change. The angular distribution of diffracted radiation has the form of individual diffraction spots which are referred to as lauegrams.

The Debye-Scherrer method (powder method) is used to study polycrystals and their mixtures. The sample is irradiated with a monochromatic X-ray beam, for example, copper with a wavelength of 1.5418 Å. The random orientation of the crystals in the sample relative to the direction of X-ray incidence transforms the diffracted beams into a family of coaxial cones with the incident beam as the axis (debyegram).

The main task of X-ray Structure Analysis (XSA) is to reveal the crystalline structures of substance, with the help of diffraction patterns, obtained in the form of a lauegram and a debyegram. The most widely used is the XSA method, developed in 1916–1917 by Peter Debye and Paul Scherrer which is largely known as a powder method. The development of this method was a significant experimental simplification compared to the Laue method, which allowed one to determine the structure only by X-ray diffraction on single crystals. In earlier instruments to record diffracted X-ray beams, forming a system of coaxial cones, a photographic film was used. The debyegram, formed by lines of X-ray cones intersecting with the film, looks like a series of slightly curved lines. The location and intensity of these lines makes it possible to judge the crystal structure and phase composition of the crystalline substance. In modern devices, another principle of recording diffraction maxima or reflexes is used, which will be considered below.

The task of X-ray Phase Analysis (XPA) is more modest and consists in recognizing unknown phases by comparing the diffraction pattern of the test substance with the diffraction patterns of known phases. Substances of different chemical composition, or of one composition, but of a different structure, have different crystal lattices and therefore give different diffraction patterns. The fundamental difference between XPA and most chemical methods of phase analysis is that it is not the elemental composition that is determined, but in which the thermodynamic phases the substances are in. In addition, the XPA method, as well as the XSA method, in contrast to chemical analysis, is non-destructive.

The main disadvantage of XPA is the relatively low sensitivity of the method. To detect the crystalline phase, its content should be at least 1% for the vast majority of substances. Significant difficulties, in addition, arise when working with amorphous phases, but even in this case, the method is sometimes very informative. This is especially when the processes of transitions from the amorphous phase to the crystalline phase are studied. Some difficulties also appear in those rare cases when substances of different compositions have geometrically identical lattices and therefore give very similar diffraction patterns. As an example, one can cite the diffraction patterns of aluminum, silver and gold, metals having the same face-centered cubic lattice with parameters differing by less than 1%. Here the complete picture may be obtained by the combination of X-ray phase analysis with elemental (for example, X-ray dispersive microanalysis).

4.3.1   X-ray analysis

The method of X-ray structural analysis, developed by P. Debye and P. Scherrer, is still the most common method for structure determining due to its simplicity, versatility and relative cheapness. The method of P. Debye and P. Scherrer is based on the phenomenon of X-ray diffraction on a three-dimensional crystal lattice. The explanation of this phenomenon was given in 1913 independently by W. Bragg and G. Wulf [9, 10].

Bragg Theory

As was shown by W. Bragg, the diffraction phenomenon can be explained if one considers the crystal as a set of atomic planes shown in Figure 4.36. X-rays with wavelength λ, penetrating the crystal, are partially reflected from each plane. The figure shows the reflection of rays from the first, second and third atomic planes. Dashed lines show the wave fronts of the incident and reflected waves. Obviously, the beam reflected from the second plane travels an additional distance Δx, which is called the path-length difference. In this case, the path difference is 2d sinΘ, where d is the interplanar distance, Θ is the angle between the direction of the X-rays and the atomic plane. The path-length difference for the beam reflected from the third plane will be 4d sin Θ, etc.

Let us characterize the oscillation amplitudes of the reflected X-ray waves as A_1, A_2, A_3,..., where $A_1 > A_2 > A_3 > ...$, and for the oscillation phases introduce the following notation $\vartheta_1 = \omega t - kx$, $\vartheta_2 = \omega t - kx - 2\pi n$, $\vartheta_3 = \omega t - kx - 4\pi n$ ... If the minimum

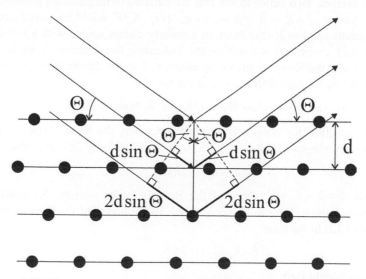

Figure 4.37. The formation of the pass-length difference of X-rays when reflected from atomic planes

value of the path-length difference equals an integer number of wavelengths λ, i.e., $\Delta x = 2d\sin\Theta = n\lambda$ then $\vartheta_2 = \omega t - kx - 2\pi n$, $\vartheta_3 = \omega t - kx - 4\pi n$ ... In accordance with the principle of superposition, the sum of these waves will be equal to

$$A_1 e^{i\vartheta_1} + A_2 e^{i\vartheta_2} + ... = A_1 e^{i\vartheta_1} + A_2 e^{i\vartheta_1} e^{-i2\pi n} + ... = (A_1 + A_2 + ...)e^{i\vartheta_1}. \qquad (4.8)$$

If the path-length difference contains an integer of half waves $\Delta x = n\lambda/2$, then formula (4.8) acquires the form

$$A_1 e^{i\vartheta_1} + A_2 e^{i\vartheta_2} = A_1 e^{i\vartheta_1} + A_2 e^{i\vartheta_1} e^{-i\pi n} + A_3 e^{i\vartheta_1} e^{-i2\pi n}$$
$$= (A_1 - A_2 + A_3 - ...)e^{i\vartheta_1} \qquad (4.9)$$

Thus if in the first case the oscillation amplitudes add up, in the second they are subtracted in pairs. Accordingly, an increase in intensity will be observed for X-ray scattered at an angle

$$\Theta = \arcsin\left(\frac{n\lambda}{2d}\right). \qquad (4.10)$$

It should be noted that the above explanation of interference has several disadvantages, because it operates representations from geometric optics along with the concepts of wave optics. For example without resorting to the Huygens-Fresnel principle, it is difficult to explain how a beam of reflected parallel rays can interfere.

A more rigorous explanation of interference was given by Laue. The difference between Laue's approach and Bragg's theory is that it does not require breaking the crystal into atomic planes. It is assumed that the crystal consists of identical atoms located at the sites of the Bravais lattice, the position of which is determined by the vector $\vec{R} = m_1\vec{a}_1 + m_2\vec{a}_2 + m_3\vec{a}_3$, where m_1, m_2, m_3 are integers, and $\vec{a}_1, \vec{a}_2, \vec{a}_3$ are primitive vectors. Two lattice atoms that are situated in neighboring positions, such that by a vector $\vec{d} = \vec{R}_2 - \vec{R}_1 = \vec{a}_1 + \vec{a}_2 + \vec{a}_3$ (Fig. 4.38) could be considered. Each of these atoms scatters X-rays from an infinitely distant source with a wave vector $\vec{k}_1 = (2\pi/\lambda)\vec{n}_1$, where $\vec{n}_1$ is a unit vector, indicating the direction X-ray incidence. Let us consider the X-rays scattered by atoms 1, 2 in the direction $\vec{n}_2$. According to Fig. 4.38 the path-length difference of these rays is

$$\Delta x = d\cos\Theta_1 + d\cos\Theta_2 = \vec{d}\cdot\vec{n}_1 - \vec{d}\cdot\vec{n}_2 = \vec{d}\cdot(\vec{n}_1 - \vec{n}_2). \qquad (4.11)$$

The condition for constructive interference has the form $\vec{d}\cdot(\vec{n}_1 - \vec{n}_2) = m\lambda$, where m is an integer. Multiplying both parts of this ratio by $2\pi/\lambda$, one obtains

$$\vec{d}\cdot(\vec{k}_1 - \vec{k}_2) = 2\pi m. \qquad (4.12)$$

Since $\vec{d} = \vec{a}_1 + \vec{a}_2 + \vec{a}_3$ is one of the Bravais lattice vectors, the condition of constructive interference can be generalized for all nodes of the Bravais lattice by rewriting (4.12) in the form

$$\vec{R}\cdot(\vec{k}_1 - \vec{k}_2) = 2\pi m, \qquad (4.13)$$

or in an equivalent form

$$e^{-i\vec{R}\cdot(\vec{k}_2 - \vec{k}_1)} = 1. \qquad (4.14)$$

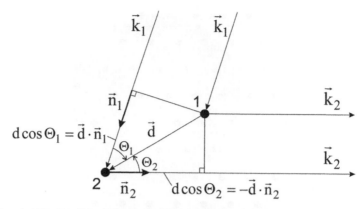

Figure 4.38. Condition for constructive interference of X-rays scattered by the atoms of the crystal lattice

Comparing (4.14) with the definition of the reciprocal lattice $e^{i\vec{R}\cdot\vec{K}} = 1$, one obtains the relation $\vec{k}_2 - \vec{k}_1 = \vec{K}$, where $\vec{K} = m_1\vec{b}_1 + m_2\vec{b}_2 + m_3\vec{b}_3$ is the reciprocal lattice vector. Thus, the condition of constructive interference is satisfied for those scattered X-rays which wave vectors k_2 satisfies the condition

$$\vec{k}_2 - \vec{k}_1 = \vec{K}. \tag{4.15}$$

This condition makes the essence of the method of measuring the parameters of the crystal lattice. It is necessary to clarify that, deriving Equation (4.15) from Equation (4.14), one can deliberately change the signs of the vectors $\vec{k}_1$ and $\vec{k}_2$ so, that further geometric constructions look more comprehensive. For example, the sphere in Fig. 4.39 looks more natural, if it is outlined by the terminal point of the vector k, rather than by its initial point.

Ewald's sphere and Laue method

From a parallel beam of incident X-rays, one can be choose the ray that falls into one of the points of the reciprocal lattice of the crystal (Fig. 4.39). Then the terminal point of the wave vector $\vec{k}_1$ of the accompanying X-ray can be placed in the selected point (site) and the direction of the wave vector of the scattered X-ray $\vec{k}_2$, corresponding the condition of constructive interference, can be defined. Since the Laue theory assumes that the X-ray scattering is elastic, i.e., with the energy and momentum magnitude conservation, the wave vectors of the incident and scattered X-rays should be equal in modulus $k_1 = k_2$. But whereas the direction of the vector $\vec{k}_1$ is determined by the experimental conditions, the direction of the vector $\vec{k}_2$ may be an arbitrary one.

In other words, the end of the vector $\vec{k}_2$ forms a locus of points equidistant from a given point '0', called the center. This locus is referred to as the Ewald sphere (Fig. 4.39). As can be seen from the figure, the condition of constructive interference (4.15) will be satisfied only if any of the points of the reciprocal lattice also belongs to the Ewald sphere. Vectors directed to these points determine the directions in which interferential maxima are observed. Knowing the vector $\vec{k}_1$ and determining

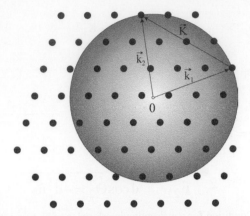

Figure 4.39. Graphic illustration of the condition of constructive interference
of X-rays scattered by the atoms of the crystal lattice

the direction of the vector $\vec{k}_2$, one can find the reciprocal lattice vector $\vec{K} = \vec{k}_2 - \vec{k}_1$
as well as determine its primitive vectors. These vectors are associated with the
primitive vectors of the direct lattice by the relations $\vec{b}_1 = 2\pi(\vec{a}_2 \times \vec{a}_3)/\vec{a}_1 \cdot (\vec{a}_2 \times \vec{a}_3)$,
$\vec{b}_2 = 2\pi(\vec{a}_3 \times \vec{a}_1)/\vec{a}_2 \cdot (\vec{a}_3 \times \vec{a}_1)$, $\vec{b}_3 = 2\pi(\vec{a}_1 \times \vec{a}_2)/\vec{a}_3 \cdot (\vec{a}_1 \times \vec{a}_2)$. The experimental
determination of the positions of interferential maxima forms the basis of the method
developed by M. Laue for determining the lattice parameters of single crystals.

The Laue method is applied at the first stage of studying the atomic structure of
crystals and is used to determine the syngony and the Laue class of a crystal. This
method is applicable only for the study of single crystals or coarse-grained samples.
Therefore, it is unsuitable for an experimental study of the nanocrystalline structure
of materials used in all-solid-state lithium-ion batteries. However, it allows a good
understanding of the essence of the Debye–Scherer method, which is widely used for
X-ray structure and X-ray phase analysis of the functional layers of SSLIB.

The essence of the method is to ensure the practical implementation of the
constructive interference conditions, since with a monochromatic X-ray source, the
probability of the Ewald's sphere to be hit by the point of the reciprocal lattice is very
small. To moderate the condition (4.15), in the Laue method a non-monochromatic
radiation whose modules of the wave vectors $\vec{k}$ belong to the interval from k_0 to k_1,
i.e., $k \in [k_0, k_1]$ is used.

Figure 4.40 shows two Ewald's spheres whose centers are located at the points
0 and 0′ at the boundaries of this interval. Now the interference condition (4.15) will
be satisfied if the reciprocal lattice site falls into the region bounded by the surfaces
of two Ewald's spheres. For each point falling into this region, one can plot a vector
$\vec{k}_2$ by choosing the initial point of the vector so, that the condition $k_2 = k \in [k_0, k_1]$
is satisfied.

To study the atomic structure of crystals, along with the Laue method, the
rotating crystal method is used. The essence of the method is that when the crystal
rotates around any of its axes, the reciprocal lattice also rotates. In this case, the
points of the reciprocal lattice move in circles, some of which intersect the Ewald's
sphere, to ensure fulfilling the condition of constructive interference.

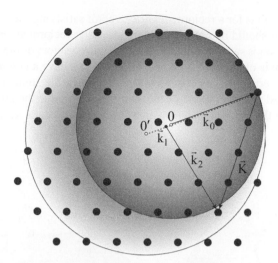

Figure 4.40. Laue method. Constructive interference is achieved through the use of non-monochromatic radiation. The modules of the X-ray wave vectors $\vec{k}$ belong to the span of values from k_0 to k_1. For all points of the reciprocal lattice, falling in the region between the Ewald's spheres of radii k_0 and k_1 the condition $K = k_2 - k_1$ holds

Debye-Scherer Method

The most practically significant method of the X-ray structure and phase analysis is the Debye-Scherer method. According to this method the test substance, if in a solid state of aggregation, it is ground in a mortar or in a ball mill. This is necessary so that the crystallites in the sample should have a chaotic orientation. Due to this feature the Debye-Scherer method is often called the powder method. The method itself in many respects is similar to the rotating crystal method, but if in the rotating crystal technique, the condition of constructive interference is fulfilled sequentially for some of reciprocal lattice vectors, then in the Debye-Scherer method this condition is satisfied for the set of such vectors, belonging to different crystallites. The fact is that crystallites are already oriented in all possible directions, therefore, the condition of the interference maximum is met by not one wave vector $\vec{k}_2$, but by a cluster of vectors, tightly filling a conical surface in the k-space. Ewald's construction for the powder method can be examined in more detail.

Since the crystallites in the sample now have a random orientation, the reciprocal lattice vector can also have an arbitrary orientation. Further, the vectors $\vec{K}$ are considered, for which condition $K < 2k$ is satisfied, where k is the modulus of the wave vector of the incident X-ray. A set of crystallites of the sample generates the set of vectors whose ends hit a surface of a sphere of radius K. If the accompanying X-ray wave vector as before is fixed then the scattered radiation wave vector forms an Ewald's sphere which intersects a sphere of radius K. The intersection line of these two spheres is a circle that is notable for the fact that for any of its points, the condition of constructive interference is satisfied. Thus, an X-ray that satisfies the condition of the interference maximum is scattered along the conical surface, as shown in Fig. 4.41a.

The Ewald's plots for a reciprocal lattice vector satisfying the condition K < 2k was considered. It should be noted that there are other reciprocal lattice vectors for which this condition is satisfied. In this case, the wave vectors of the scattered X-ray form cones, the axis of which is a vector $\vec{k}_1$. The cones have a common vertex, and the angles at the vertices of the cones will be the larger, the smaller is the modulus of the vector K. At a certain value of the vector $\vec{K}$ modulus the cones 'turn inside out', as shown in the Fig. 4.41b, and with a further decrease in K, it converges to a vector $\vec{k}_1$.

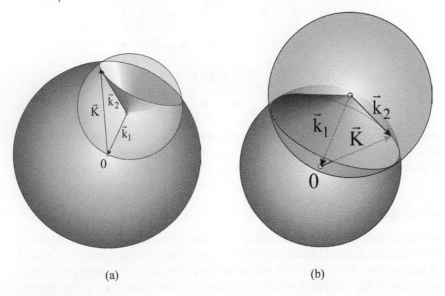

(a)   (b)

Figure 4.41. Debye-Scherer method. A set of randomly oriented crystallites of the sample, generates a set of randomly oriented reciprocal lattices. The ends of the reciprocal lattice vectors under condition K = const belong to a sphere of radius K centered at point 0. Two further cases are possible: a). The cone of scattered X-ray is directed towards the X-ray source. b). The cone of scattered X-ray is directed to the side opposite to the X-ray source

It should be clarified that due to the finiteness of the crystallites number, the set of vectors $\vec{K}$, where K = const will also be finite, and the number of points on the surface of a sphere of radius K will be finite as well. Therefore, it is possible that the circle along which two spheres intersect will pass through 'empty places', or scattered X-ray wave vectors will be directed away from the radiation detector. This probability is partly removed by the fact that the X-ray line (for example, Cu – K_α) has a finite width. If for the width $\Delta\lambda$ of the line K_α is taken its Full Width at Half Maximum (FWHM), then the modulus of vector $\vec{k}_1$ will fall in the interval from $k_1' = 2\pi/(\lambda + \Delta\lambda/2)$ to $k_1'' = 2\pi/(\lambda - \Delta\lambda/2)$. The Ewald's spheres of the corresponding radii will intersect the sphere of radius K along two circles, as shown in Fig. 4.42. In this case, in order for the condition of constructive interference to be fulfilled, the end of the vector $\vec{K}$ of a crystallite should not fall strictly on the circle, but in the ball belt enclosed between these circles. Thus, increases in the probability for the points of reciprocal lattice to fall into this zone. It can be seen that the Debye-

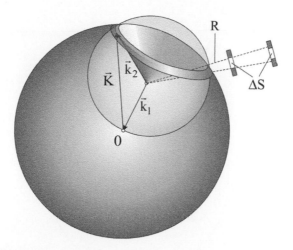

Figure 4.42. With the finite width of $Cu - K_\alpha$ line of the X-ray source, the Ewald's spheres of radius $k_1' = 2\pi/(\lambda + \Delta\lambda/2)$ and $k_1'' = 2\pi/(\lambda - \Delta\lambda/2)$ cut out on the surface of a sphere of radius K a ball belt. The figure shows only the Ewald's sphere corresponding to the center of the $Cu-K_\alpha$ line. If the distance from the sample to the detector increases, the flux of X-rays incident on aperture of the area Δs decreases, while the angular resolution of the device improves

Scherer method combines both elements of the Laue method and the method of a rotating crystal.

In real systems, the X-ray source moves continuously, so that the end of the vector $\vec{k}_1$ rotates around the center of the Ewald's sphere. In this case, the detector 'sees' reflexes, the wave vectors of which are enclosed within a finite solid angle $\Delta\Theta = \Delta S/R^2$, where R is the radius of the circle along which the source and detector of the X-ray radiation move, and $\Delta S = \Delta l \cdot \Delta h$ is the area of the detector's aperture, where Δl and Δh are the widths of the vertical and horizontal slits of the input collimator.

If one positions the vectors $\vec{k}_1$ and $\vec{k}_2$, lying in the picture plane, as shown in Fig. 4.43, and lower the normal N to the plane of the sample at the point of the X-ray incidence then from the Ewald's constructions it is obvious that the vectors $\vec{k}_1$ and $\vec{k}_2$ should be located at the same angle relative to the normal. In this case, a set of different atomic planes perpendicular to N 'work' to reflect the X ray in the direction of the detector. X-ray source 1 and detector 2 move towards each other with the same angular velocities. It can be assumed that at a certain angle 2Θ the condition of constructive interference is fulfilled. As can be seen from Fig. 4.43, the modulus of the reciprocal lattice vector is related to the angle Θ by the equation

$$K = 2k_1 \sin\Theta. \tag{4.16}$$

Moreover, it is perpendicular to the atomic planes for which the condition of constructive interference is fulfilled. The reciprocal lattice vector perpendicular to the atomic plane (hkl), where h, k and l are the Miller indices defining this plane and the set of parallel planes, can be represented as $\vec{K}_{hkl} = h\vec{b}_1 + k\vec{b}_2 + l\vec{b}_3$. The modulus of such a vector is related to the interplanar distance d_{hkl} as $K_{hkl} = 2\pi/d_{hkl}$. In turn, the vector $\vec{K}$ can be associated with $\vec{K}_{hkl}$ in the only possible way $\vec{K} = n\vec{K}_{hkl}$, where n

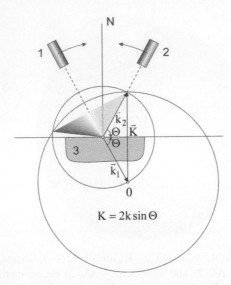

Figure 4.43. Ewald's construction for the Debye-Scherer method, performed in the plane of the goniometer. Arrows indicate the direction of motion of the source and X-ray detector

= 1,2,... . Then the modulus of the vector K is $K = 2\pi n/d_{hkl}$ and formula (4.16) can be written as

$$d_{hkl} = \frac{n\lambda}{2\sin\Theta}.$$ (4.17)

The resulting expression is identical to the condition of constructive Wulf-Bragg interference (4.10). Thus, by measuring the angles at which interference maxima are observed, it is theoretically possible to find the ratio d/n, where n = 1,2,... and determine the interplanar distance d. As a rule, the wavelength and interplanar distance are of the same order of magnitude. Therefore, by virtue of the relation $\sin\Theta = n\lambda/2d$ and appearing from its condition $0 < n\lambda/2d < 1$, for most of crystal lattices in the X-ray diffraction pattern there cannot be more than one reflex, satisfying this condition. For gratings with large interplanar distance, second and third order reflexes can be observed. For most materials used in lithium-ion batteries, we can assume that n = 1.

Based on the obtained set of interplanar distances, it is possible to 'manually' establish the type and parameters of the crystal lattice. If, however, several phases are present in the sample, then the task becomes much more complicated. Therefore, in practice for X-ray structure analysis and X-ray phase analysis databases, containing X-ray diffraction patterns of all known substances having a crystalline structure are used. The most widely used is The Powder Diffraction File (PDF) of The International Center for Diffraction Data (ICDD) which collects, edits, publishes and distributes powder diffraction data for the identification of materials. To search the databases and compare the results with the diffraction patterns, contained in the databases, special software is used that is supplied with the databases.

4.3.2   Implementation of the Debye-Scherer method

Hardware implementation

As can be seen from Fig. 4.43, the essence of the Debye-Scherrer method consists in irradiating a polycrystalline sample with monochromatic X-ray and measuring the angles at which the condition of constructive interference is satisfied. Therefore, the main part of the X-ray diffractometer is a goniometer, on which the source and detector of X-ray radiation are located. Most modern goniometers are of the vertical $\Theta - \Theta$ type, which is more preferable than the traditional horizontal. The main reason is that the stage in these goniometers is horizontal and does not move during scanning, which allows even bulk powders or liquids to be analyzed. Bragg-Brentano focusing is used when the radiation source and detector move along a circle of R_G radius in the center of which a flat sample is placed. The collimator of the X-ray source and the entrance slit of the detector are located on the so-called focusing circle of radius $R_F = R_G/2\sin\Theta$ (Fig. 4.44). This ratio can be easily obtained as the difference of two segments of length $R_G\sin\Theta$, and $R_F\cos2\Theta$, depicted in Fig. 4.44. Subtracting the short segment from a long one, one gets the ratio

$$R_F = R_G\sin\Theta + R_F\cos2\Theta, \tag{4.18}$$

from which the relation $R_F = R_G/2\sin\Theta$.

Figure 4.44 shows that a diverging X-ray beam, reflected from the focusing circle, is collected at the entrance slit of the detector. For this, it is necessary that

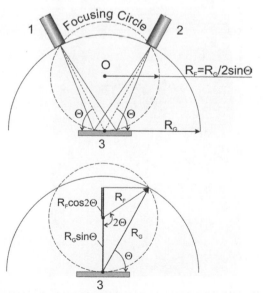

Figure 4.44. Bragg–Brentano focusing: 1 is X-ray tube, 2 is detector, 3 is sample, O is center of the focusing circle. The lower part of the figure shows the constructions explaining the relation between radii R_G and R_F

the sample's surface coincides with the surface of the focusing circle, the curvature of which depends on the angle of X-ray incidence Θ. In practice this, of course, is impossible, and the focusing condition is fulfilled only approximately, i.e., the plane surface of the sample only touches the focusing circle. In a course of measurement, the X-ray tube and detector move along the circumference of the goniometer with equal angular velocities in opposite directions. Their rotation is carried out using a stepper motor, which provides a stop of the sweep at certain angles of Θ, for a certain time span to count the number of X-ray quanta.

As an example, an ARL X'tra X-ray diffractometer (Fig. 4.45) can be considered. Most powder diffractometers currently use the Bragg-Brentano vertical focus option $\Theta - \Theta$. In this geometry, the ARL X'tra is optimally configured to allow upgrades or reconfigurations with optional accessories. The main element in the design of the device is a goniometer with two moving alidades that synchronously rotate around a common center (Fig. 4.46). The device allows one to move source 4 and detector 3 along the alidade, changing the diameter of the goniometer D in the range of 400–520 mm. In a standard setting, the goniometer diameter is 520 mm. As can be seen in Fig. 4.43, adjustment of the goniometer diameter and slit width of the input collimator allows one to change the angular resolution of the device.

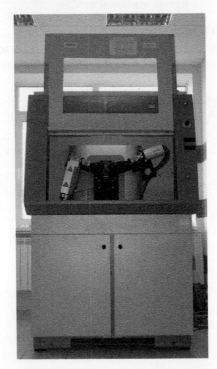

Figure 4.45. ARL X'tra X-ray powder diffractometer implementing
the Debye-Scherer method

The range over which the 2θ angle can vary in the ARL X'tra diffractometer is from –8° to 164°. The range may be less if the goniometer diameter is less than

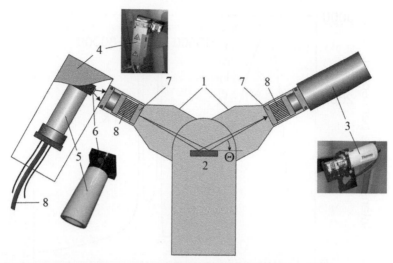

Figure 4.46. Schematic of X-ray diffractometer: 1 are goniometer alidides, 2 is a stage for sample, 3 is X-ray detector unit, 4 is X-ray source block, 5 is ceramic X-ray tube, 6 is beryllium window, 7 are primary and diffracted X-ray collimators, 8 are Soller slits

520 mm and/or if some attachments are installed. The diffractometer is equipped with a digital high-precision stepper motor with an optical decoder, the positioning accuracy of which is ± 0.00025 degrees.

The X-ray detector unit 3 is mounted directly on the goniometer and provides maximum stability and ease of adjustment, as well as the possibility of working in the line or point focus mode. The collimators of the primary and diffracted beams 7 have two collimator slits – horizontal and vertical, located on both sides of the Soller slits 8. The slits are equipped with micrometer screws to adjust the width of the slit in the range from 0 to 10 mm. The slits limit the axial (in the picture plane) beam deviations within ±1.3°.

The scattered X-ray radiation is registered using a semiconductor detector Si(Li), which is mounted on a 5-stage water cooled Peltier module. The detector has a very high efficiency and enables detecting almost 100% of X-ray photons. An important feature that allows correct interpretation of diffraction patterns is its high linearity. For the Si(Li) detector, the dose-response function remains linear at least up to 60,000 pulses per second, and the internal background is below 0.1 pulses per second. The energy resolution of the detector is up to 250 eV. In addition, a quick retuning of the energy window and a choice of K_α (doublet) or K_β (singlet) is possible.

The source of X-ray radiation is a ceramic X-ray tube 5 with a cooled anode. At the base of the tube there are four beryllium windows 6 for the output of X-ray radiation and openings for supplying water cooling. The tube is placed in a holder in which an electromagnetic shutter is provided to protect against X-ray radiation.

As noted above, when braking electrons with energy of 60 kV in the massive anode, X-ray bremsstrahlung and characteristic X-ray emission is generated. The spectrum of such an X-ray is shown in Fig. 4.47. As can be seen from the figure, there is a certain minimum wavelength λ_{min}, which is determined by the obvious relation

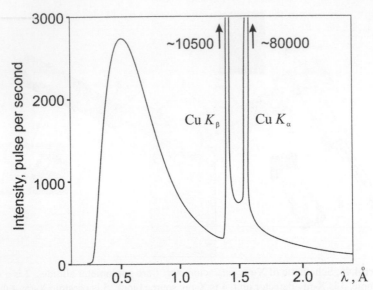

Figure 4.47. The emission spectrum of an X-ray tube with a copper anode

$eU \geq h\nu$ followed from energy conservation law. The latter means that the energy of the X-ray quantum $h\nu$ cannot exceed eU – the energy of the electron exciting this quantum, where U is the potential difference between the anode and cathode of the X-ray tube. Using the relation $\lambda\nu = c$, it is easy to find the minimum wavelength λ_{min} = hc/eU.

Above this wavelength, the bremsstrahlung intensity increases, reaching a maximum value, and then tends to zero. Against the background of bremsstrahlung, sharp peaks of anode radiation are observed. The area of these peaks is about 1% of the area under the bremsstrahlung curve. The complex spectral composition of X-ray tube radiation makes it difficult to perform research on metals and alloys. To attenuate unwanted lines of characteristic X-ray and brake radiation various filters are used. Such filters selectively absorb one component of the spectrum and pass the other with minimal attenuation. To do this, the so-called K-jumps of absorption are used. The absorption spectrum of such a filter and filtered x-ray spectrum can be seen in Figs. 4.48, 4.49, provided that the filter substance has a K-jump between the lines of the spectrum.

This condition is met by a chemical element whose atomic number is one less than the serial number of the metal of the anode of the X-ray tube. For example, to remove a component of the characteristic spectrum of an X-ray tube with an iron anode (Z = 26), one can select the chemical element number 25 as a filter, i.e., manganese which is the nearest neighbor of iron in the periodic system. Similarly, for the copper anode (Z = 29), the filter can be nickel (Z = 28), and for molybdenum (Z = 42) it can be zirconium (Z = 41). Radiation filters are thin metal foils or powder oxide composites deposited on cardboard. Since the attenuation of the intensity of the component K_β can be in the range of 50-500 times, and the component K_α in 2-3 times, the thickness of the layer of metal or powder is essential for filtering radiation. Usually the diffractometer is equipped with a copper anode, so a nickel beta filter

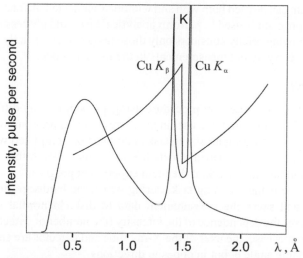

Figure 4.48. The position of the K–jump of the absorbing substance for K_β filter (beta filter) is between K_α and K_β copper peaks

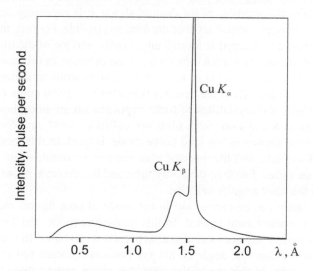

Figure 4.49. X-ray spectrum of copper after the beta filter

is included in the package, although installation of Mn, V, Fe and Zr filters is also possible.

Software implementation of the powder method

The standard program for X-ray diffractometer operation consists of four main parts:
- hardware configuration;
- manual start of hardware (diffractometer menu);
- data acquisition;
- data analysis.

The program can be used both with and without a connected diffractometer. In the latter case, the program is used to perform analytical tasks and process measurement results. The following briefly considers only those tasks that are usually supported by the program for carrying out measurements in the manual mode.

Data acquisition

The first step in the measurement procedure using a modern X-ray diffractometer is to collect data. To do this, first, the user in the corresponding program window forms a list of tasks. When creating a list, the tasks that must be completed when working on the device, should be described in detail. Each of the tasks in this sequence is a subroutine within the framework of the main experiment program.

The main one is the scanning task, during which the hardware runs in a given angular range and saves the measurement data to disk. In normal scanning, the program measures the dependence of the intensity (the number of detector counts) on the angle 2θ. In this case, the axes of the X-ray tube and detector are simultaneously rotated by the same angle θ, but in opposite directions.

In the study of thin films (less than 5 μm), the signal from the film itself will always be weak, and at the same time, it can be superimposed by the signal from the substrate. To raise the intensity of the desired signal, it is necessary that the X-ray radiation passes through as thick layer of the film, as possible. For this, the axis of the X-ray source should be located at a small angle to the surface of the film, but at the same time the radiation should not directly enter the detector. In this scanning mode, the axis of the X-ray tube is set at an angle $\theta < 5°$ and remains stationary during the measurement, while the axis of the detector is rotated in a given range of angles.

In some cases, the crystallites of films deposited on an amorphous substrate, have a preferential orientation. Such films are called textured, and for their study, a scanning method known as 'rocking curve mode' is used. In this mode, the angle between the X-ray tube and the detector axes remains constant, while the angle of X-ray incidence varies. For this, the X-ray tube and the detector rotate in the same direction with the same angular velocity.

A texture study can be carried out in the mode of pole figures measuring, for which the texture attachment is used. In this measuring mode, the X-ray tube and detector are stationary and are located symmetrically with respect to the vertical axis of the goniometer; thus, the angle 2θ always remains constant towards the initial position of the stage. In this mode, the stage can rotate around its vertical axis by $360°$ and tilt relative to the θ-θ angles' plane by an angle from $-5°$ to $90°$.

All the listed scanning modes allow performing qualitative and quantitative analysis, determining interplanar distances, types of crystal lattices, crystallite sizes and identify atomic planes of textured films.

Data analysis

At the end of data acquisition and their saving in the source file, their analysis is carried out. During the analysis process, which proceeds in several stages, all data stored in the source file are processed. Firstly, the background appearing from the diffuse scattering of X-rays is subtracted, and as a result the diffraction pattern becomes clearer. Then, the diffraction pattern is smoothened and adjusted. The peak search program is started, which saves the data on the peaks found onto a file. During the last stage of data analysis, a profiling program is carried out, which generates information about the exact position of the peaks, their shape and intensity. Processed in this way, the measurement results contain information about the sample, such as the name of the phases, their number, crystallinity, crystallite size, preferred orientation (in case of texture detection), level of residual stress, structural data and crystal lattice parameters. Below are some examples of the diffraction patterns analysis for materials of solid-state lithium-ion batteries obtained by an ARL X'tra diffractometer.

4.3.3 X-ray phase analysis of materials for all-solid-state lithium-ion batteries

Silicon nanocomposite films

One of the most promising materials for the negative SSLIB electrode is a Si–O–Al nanocomposite, the main part of which is amorphous silicon [11-15]. The specific capacity of such a material depends on the elemental composition and in liquid electrolytes they can vary over a wide range – from several hundreds to 3000 mA · h/g. Si-O-Al films are deposited by magnetron sputtering of silicon and aluminum targets in a nitrogen plasma under controlled oxygen inflow. A typical Si-O-Al film composition is given in Table 4.4. The main purpose of oxygen and aluminum additives is to stabilize the amorphous state of silicon and increase the electronic conductivity of the film. Partial oxidation and disordering of the silicon structure prevents its crystallization with the formation of a metastable phase $Li_{15}Si_4$. Therefore, the main task in developing the technology of film deposition is to obtain precisely amorphous silicon, and the exact and available method to control the structure and phase composition of the films is X-ray phase analysis. Figure 4.50 shows the diffraction patterns of polycrystalline and amorphous Si-O-Al films deposited on a silicon wafer.

Table 4.4. Parameters of a Si-O-Al film

Film parameters	
Film thickness, μm	1,34
Resistivity, Ohm·cm	0,74
Concentration C, at. %	4,94
Concentration O, at. %	31,91
Concentration Al, at. %	8,76
Concentration Si, at. %	53,88
Concentration Ti, at. %	0,51
Total, at. %	100

In the course of working out technology for Si-O-Al films magnetron deposition, it is often necessary to know how the structure of the film affects its capacitive

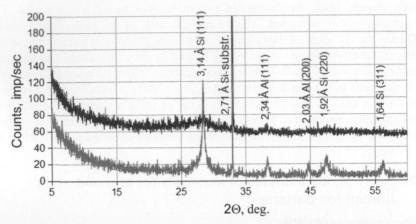

Figure 4.50. Diffraction patterns of Si-O-Al films: polycrystalline film (lower curve) and amorphous (upper curve). The interplanar distances and Miller indices of the atomic planes at which X-ray interference takes place are indicated. 'Si-substr' is reflex from a silicon wafer

characteristics and cycleability. For this, the charge-discharge characteristics of the film are tested in laboratory half-cells, where a lithium foil acts as a counter electrode. The electrode and counter electrode are separated by a non-woven polypropylene separator, which is impregnated with electrolyte LP-71 (1M solution of $LiPF_6$ in a mixture of ethylene carbonate (EC), diethyl carbonate (DEC), dimethyl carbonate (DMC) (1:1:1)). In the process of measuring charge-discharge characteristics, numbering up to a thousand charge-discharge cycles and more, electrolyte degradation occurs with the formation of solid lithium compounds. The method of the X-ray phase analysis allows to explore these processes as well. Figure 4.51 shows the diffractogram of the separator after cycling, in which lithium fluoride crystallites was formed.

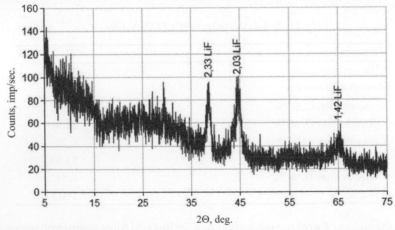

Figure 4.51. The diffractogram of the separator after cycling. The interplanar distances of lithium fluoride crystals are indicated

Nanostructured Vanadium Oxides

A number of vanadium oxides are of considerable interest to developers of lithium-ion batteries due to their ability to reversible lithium insertion in large quantities. Of these, the most promising material is vanadium pentoxide, the theoretical specific capacity of which is 883 mA · h/g. This is significantly higher than the capacity of traditionally used materials, such as, $LiCoO_2$, $LiFePO_4$, $LiMn_2O_4$ and others. However, the process of controlled formation of vanadium oxides films of the required composition is complicated by the fact that about 25 compounds are present on the phase diagram of a two-component vanadium–oxygen system [16]. Of these, in films obtained by magnetron sputtering, often the most observed are:

- V_2O, $VO_{0.9} - VO_{1.3}$, VO_2;
- V_2O_5, V_3O_7, V_6O_{13} general formula V_nO_{2n+1};
- V_2O_3, V_3O_5, V_4O_7, V_5O_9, V_6O_{11}, V_7O_{13}, V_8O_{15} general formula V_nO_{2n-1}.

The technology for the manufacture of vanadium oxide films provides for their oxygen annealing in the temperature range in which beside the above compounds, solid solutions of oxygen in vanadium can form, which considerably complicates the study of such systems. Therefore, the phase composition diagnostics during the development of technology for films magnetron deposition is an urgent task. An example of diffraction patterns of films deposited on a silicon wafer and titanium foil is shown in Fig. 4.52.

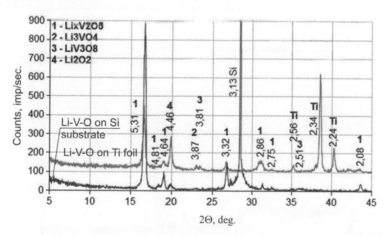

Figure 4.52. Diffraction patterns of films on a single-crystal silicon wafer (lower curve) and on a titanium foil (upper curve). The interplanar distances and Miller indices of atomic planes are indicated. Reflexes from silicon and titanium are marked with Si and Ti. Reflex numbers correspond to the phases indicated in the upper left part of the figure

Solid electrolyte LiPON

In most of the industrial SSLIB as a solid electrolyte lithium phosphorus oxynitride (LiPON), first synthesized by the Oak Ridge National Laboratory (Ch. 1, ref. [6]), is used. LiPON is deposited by magnetron sputtering of lithium orthophosphate Li_3

PO_4 in a nitrogen-argon plasma, resulting in the compound $Li_3PO_xN_y$. To achieve an ionic conductivity of the order of 10^{-6} S/cm, a LiPON film must contain at least 7% nitrogen and have an amorphous structure. The percentage of nitrogen and other elements is measured by X-ray dispersive microanalysis, while the amorphous structure of the electrolyte is usually tested by X-ray diffractometry methods. Table 4.5 shows the results of a LiPON analysis by means of an Apollo X energy dispersive analyzer (the attachment to Quanta 3D 200i SEM). Figures 4.53 and 4.54 show the diffraction patterns of LiPON films on different substrates obtained on an ARL X'tra powder X-ray diffractometer.

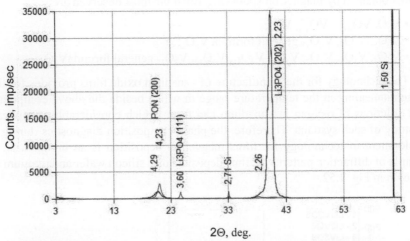

Figure 4.53. Diffraction patterns of textured LiPON films on silicon wafer

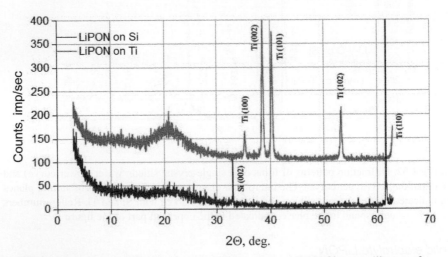

Figure 4.54. X-ray diffraction patterns of amorphous LiPON films on silicon wafer (bottom curve) and on titanium foil (upper curve). The peaks in both curves belong to the substrates, while the amorphous LiPON gives a small halo in the region of peaks $2\Theta = 22.34°$ and $2\Theta = 23.19°$ (Fig. 4.53)

These diffraction patterns belong to different stages of technology development for the LiPON magnetron deposition, therefore, they have significant differences. Initially, to prevent overheating and crystallization of LiPON, films were deposited layer by layer at a magnetron power of 200 W, with a floating bias and making small intervals between the layer's deposition. In this case, the films proved to be textured, as evidenced by the diffraction pattern shown in 4.53. There is no halo, inherent to amorphous structure and the intensity of the lithium orthophosphate peaks depends on the accuracy of the sample setting. Subsequently, magnetron deposition regimes were selected that made it possible to obtain X-ray amorphous LiPON films.

Table 4.5. Typical element concentrations in a LiPON film

Element-characteristic X-ray	Accelerating voltage			
	5 kV		7 kV	
	Element concentration, weight. %	Element concentration, at. %	Element concentration, weight. %	Element concentration, at. %
N-K_α	12,16 ± 0,63	16,79	11,85 ± 0,53	16,5
O-K_α	48,58 ± 0,91	58,7	47,43 ± 0,72	57,85
P-K_α	39,26 ± 1,15	24,51	40,72 ± 0,70	25,65
Total:	100%	100%	100%	100%

Figure 4.54 shows the diffraction patterns of amorphous LiPON, deposited on a silicon wafer and titanium foil. On both diffractograms there is a halo located near the following peaks: Li_3PO_4: $2\Theta = 22.34°$, d = 3.98 Å (101) (see Fig. 4.53) and $2\Theta = 23.19°$d = 3.83 Å (021), the orthorhombic lattice, ICDD card 01-083-0338. In addition, peaks of the substrates are visible in the diffraction patterns. On the upper diffractogram, these are peaks of titanium, and on the bottom curve, the peak is of silicon. According to ICDD card 00-001-1197 peaks on the upper curve are:

- Ti (HCP), $2\Theta = 35.02°/2,56$ Å (100);
- Ti (HCP), $2\Theta = 38.44°/2,34$ Å (002);
- Ti (HCP) $2\Theta = 40.23°/2,24$ Å(101);
- Ti (HCP) $2\Theta = 52.88°/1,73$ Å(102);
- Ti (HCP) $2\Theta = 62.73°/1,48$ Å(110),

where HCP means a hexagonal close-packed lattice, and the three hexagonal lattice indices (h,k,l) correspond to the four indices (h,k,-(h+k),l).

Silicon (lower diffraction pattern) has a face-centered cubic lattice, therefore, peak $2\Theta = 33°$, d = 2.715 Å (002) is forbidden. The peak $2\Theta = 61.7°$ belongs to the (004) plane of silicon, but in the present case it is a reflex of the tungsten characteristic X-ray. The reason for this peak's appearance is that the tungsten cathode of the tube in course of time is deposited on the copper anode. Under the condition of intense X-ray diffraction on the atomic planes of silicon, a tungsten peak also appears. Therefore, the intense Si (004) peak, which is observed at slightly larger angles, also has a tungsten satellite.

4.4 Electrochemical impedance spectroscopy

4.4.1 Electrical impedance

Complex numbers

Numbers of the form $Z = a + ib$, where a and b are real numbers, and i is the imaginary number or imaginary unit, are called complex numbers. An imaginary unit is a number whose square modulus is -1, or $i^2 = -1$. The complex number may be considered as an extension of a number line into a complex plane, where parameters a and b are the coordinates of point M in Cartesian coordinates (Fig. 4.55). The position of this point M on the complex plane can be set in polar coordinates by specifying the modulus of vector $\vec{\rho}$, that represents the position of the point, and the angle φ, that the position vector forms with the X axis. Segment $\rho = \sqrt{a^2 + b^2}$ is the modulus of a complex number, and an angle $\varphi = \arctan(b/a)$ is its phase. The complex number itself is expressed through the modulus and phase as $z = \rho e^{i\varphi}$. If one uses the Euler formula $e^{i\varphi} = \cos\varphi + i\sin\varphi$, then one can easily go from polar coordinates to Cartesian $z = \rho\cos\varphi + i\rho\sin\varphi = a + ib$.

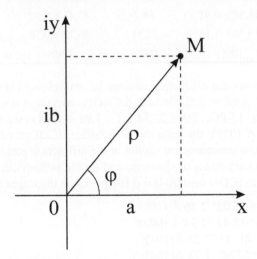

Figure 4.55. Representation of a complex number on a complex plane

Complex numbers are very convenient for the study of mathematical models of dynamical systems, since they significantly simplify the calculations. However, when interpreting the results obtained, certain problems arise. If it is about linear equations and dynamic variables, such as coordinates, speed, acceleration, then the physical meaning is attributed to the real part of solution Re(z), since its imaginary part Im(z) is discarded as having no physical meaning. Such an interpretation does not create a conflict between mathematicians and physicists, since the real and imaginary parts do not mix. Another aspect arises when it comes to non-linear dynamic variables and attempts to assign the physical meaning to the squared modulus of a complex number. From the point of view of mathematics, such an interpretation is incorrect.

As further consideration will proceed in terms of alternating current, for which the designation 'i' is traditionally used in physics and electrical engineering, then the imaginary unit will everywhere be denoted by the symbol 'j'.

Representation of electrical resistance in a complex form

Impedance is the ratio of the complex amplitude of the voltage of a harmonic signal applied to a two-terminal network to the complex amplitude of the current flowing through a two-terminal network in a steady state, that is, after the completion of transient processes

$$\hat{Z}(i,\omega) = \frac{U(\omega)e^{j(\omega t + \varphi_u)}}{I(\omega)e^{j(\omega t + \varphi_i)}} = \frac{U(\omega)e^{j\varphi_u}}{I(\omega)e^{j\varphi_i}} = \frac{\hat{U}(\omega)}{\hat{I}(\omega)}. \tag{4.19}$$

Here j is the imaginary unit, ω is the angular frequency of the harmonic voltage oscillation, t is time, φ_u and φ_i are the phases of voltage and current, respectively, $U(\omega)$ and $I(\omega)$ are the real amplitudes of the voltage and current, $\hat{U}(\omega)$ and $\hat{I}(\omega)$ are the complex amplitudes of the voltage and current. For linear passive circuits with constant parameters in a steady state, the impedance is time-independent. If the time t in the mathematical expression for the impedance cannot be reduced, then for the given two-terminal network the concept of impedance is not applicable.

Further an electric circuit (Fig. 4.56) that contains an active resistor, inductor and capacitor, to which voltage alternates according to a harmonic law, i.e., the law of sine or cosine, is considered. In its most general form, this alternating voltage can be written with the help of complex numbers as $u(t) = U_0 e^{j\omega t}$. According to second law of Kirchhoff, the sum of voltages around any closed loop is equal the sum of electromotive forces (EMF), or

$$iR + \frac{q}{C} = E + U_0 e^{j\omega t}, \tag{4.20}$$

where i is the current, R is the active resistance, q is the capacitor charge, C is the capacitor's capacitance, E is the EMF of coil, U_0 is the amplitude of an external source voltage. Given the law of induction, Equation (4.20) can be written as

$$L\frac{di}{dt} + Ri + \frac{q}{C} = U_0 e^{j\omega t}, \tag{4.21}$$

and, considering the ratio $i = \dot{q}$, reduced to the final form

$$\ddot{q} + 2\beta\dot{q} + \omega_0^2 q = u_0 e^{j\omega t}, \tag{4.22}$$

where $\ddot{q} = d^2q/dt^2$, $\dot{q} = dq/dt$, $2\beta = R/L$, $\omega_0^2 = 1/LC$, $u_0 = U_0/L$.

This equation has the same form as the equation of simple pendulum oscillating in a dissipative medium

$$m\ddot{x} + r\dot{x} + kx = F_0 e^{j\omega t}, \tag{4.23}$$

where m is the mass, r is the coefficient of medium resistance, k is the coefficient of elasticity, F_0 is the amplitude of the external force oscillations. These equations allow drawing analogs between electrical and mechanical vibrations. Comparing Equations (4.21) and (4.23), it is easy to see that the inductance is analogous to the mass and is a

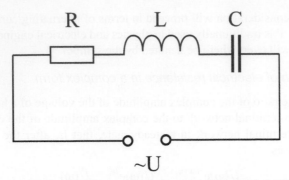

Figure 4.56. An electrical circuit consisting of resistor R, inductor L and capacitor C

measure of the electric charges' inertia. The term Ri is similar to the dissipative force, and the active resistor R is analogous to the friction factor of viscous medium r. The term q/C is analogous to a quasi-elastic force, and the reciprocal of the capacitance has the same meaning as the coefficient of elasticity. These analogs may prove to be very useful for understanding and the interpretation of the results of impedance spectroscopy.

According to the theory of ordinary differential equations, the general solution of Equation (4.22) is the sum of the general solution of the homogeneous equation

$$\ddot{q} + 2\beta\dot{q} + \omega_0^2 q = 0 \qquad (4.24)$$

and any particular solution to Equation (4.22). The general solution of the homogeneous Equation (4.24) is sought for in the form $q_1 = e^{\lambda t}$ that is substituted into Equation (4.24). The result is a quadratic equation

$$\lambda^2 + 2\beta\lambda + \omega_0^2\lambda = 0, \qquad (4.25)$$

which has two roots

$$\lambda_{1,2} = -\beta \pm \sqrt{\beta^2 - \omega_0^2}. \qquad (4.26)$$

Each root of (4.26) corresponds to a particular solution of Equation (4.24), and its general solution is a linear combination of particular solutions

$$q_1(t) = C_1 e^{\lambda_1 t} + C_2 e^{\lambda_2 t}, \qquad (4.27)$$

where C_1 and C_2, in the general case, are complex numbers.

It is easy to see that the real parts of the roots (4.26) are negative and, therefore, in the closed system ($u_0 = 0$), both particular solutions decay, i.e., tend to zero at $t \rightarrow \infty$. In order to find out how an open system evolves, a particular solution to the inhomogeneous Equation (4.22) should be found. A solution is sought for in the same form as the non-uniform term of the equation itself, namely $q_2 = q_0 e^{j\omega t}$. Substituting the desired solution in Equation (4.22) yields

$$q_0[(\omega_0^2 - \omega^2) + j2\beta\omega_0] = u_0 \qquad (4.28)$$

from which one can find the amplitude of the particular solution

$$q_0 = \frac{u_0}{(\omega_0^2 - \omega^2) + j2\beta\omega}, \qquad (4.29)$$

and then the particular solution to the Equation (4.22)

$$q_2(t) = \frac{u_0}{(\omega_0^2 - \omega^2) + j2\beta\omega_0} e^{j\omega t}. \tag{4.30}$$

The general solution of the initial Equation (4.22) is

$$q(t) = q_1(t) + q_2(t) = e^{-\beta t}\left(C_1 e^{\sqrt{\beta^2 - \omega_0^2}t} + C_2 e^{-\sqrt{\beta^2 - \omega_0^2}t}\right) + \frac{u_0}{(\omega_0^2 - \omega^2) + j2\beta\omega_0} e^{j\omega t}, \tag{4.31}$$

where the first term decays exponentially. Thus, the steady state of an electrical circuit is described by a solution

$$q(t) = \frac{u_0}{(\omega_0^2 - \omega^2) + j2\beta\omega_0} e^{j\omega t}. \tag{4.32}$$

By differentiating this expression, one can find an alternating current

$$\hat{I}(\omega) = \frac{j\omega u_0}{(\omega_0^2 - \omega^2) + j2\beta\omega} e^{j\omega t} = \frac{\hat{U}(\omega)}{L[(\omega_0^2 - \omega^2) + j2\beta\omega]/j\omega}. \tag{4.33}$$

Comparing the expression obtained with the definition of impedance (4.19), it is easy to see that

$$\hat{Z}(\omega) = \frac{L(\omega_0^2 - \omega^2) + jL2\beta\omega}{j\omega} = R + j\left(L\omega - \frac{1}{\omega C}\right). \tag{4.34}$$

Since only the observable dynamic variables have physical meaning, in this case the meaning is attributed to the complex resistance modulus, which is defined as $Z = \sqrt{\hat{Z}^*(\omega)\hat{Z}^*(\omega)}$, where $\hat{Z}^*(\omega)$ is the physical quantity complex conjugate $\hat{Z}(\omega)$. The denominator of the last fraction in the expression (4.33) can be represented in the form $Z(\omega)e^{j\varphi}$ where

$$Z(\omega) = \sqrt{R^2 + \left(L\omega - \frac{1}{\omega C}\right)^2} \tag{4.35}$$

and φ are the modulus and the phase of impedance. Then Equation (4.33) takes the form

$$\hat{I}(\omega) = \frac{U_0(\omega)e^{j(\omega t - \varphi)}}{Z(\omega)} = \frac{U_0(\omega)}{Z(\omega)}[\cos(\omega t - \varphi) + j\sin(\omega t - \varphi)]. \tag{4.36}$$

Isolation of the real part of the current in Equation (4.36) yields the expression for its amplitude as

$$I_0(\omega) = \frac{U_0(\omega)}{Z(\omega)}\cos(\omega t - \varphi). \tag{4.37}$$

Hence, the physical meaning of the impedance modulus is that it determines the amplitude value of the current in the circuit.

Impedance of perfect elements

Equation (4.34) allows in obtaining expressions for the impedance of each of the elements of the electric circuit shown in Fig. 4.56. Assuming that in this equation $L = 0$, one reaches the obvious relation

$$\hat{Z}(\omega) = R - \frac{j}{\omega C}. \tag{4.38}$$

If one also equates to zero the capacitance $C = 0$, then the impedance will be equal to infinity, because zero capacity simply means an open circuit. Therefore, in order to exclude the contribution of capacitance to impedance, let the frequency go to infinity, which results in the expression

$$Z_R = R, \tag{4.39}$$

which does not depend on frequency anymore.

The impedance of the ideal capacitor could help one obtain by setting $R = 0$ and $L = 0$

$$\hat{Z}(\omega) = R - \frac{j}{\omega C}. \tag{4.40}$$

To find the impedance of the ideal inductor, in expression (4.35) it is necessary to assume $R = 0$, and $\omega \to \infty$. Then this expression takes the form

$$\hat{Z}_L(\omega) = j\omega L. \tag{4.41}$$

Knowing the impedance of ideal elements, it is easy to calculate the impedance of any electrical circuit composed of such elements without solving differential equations. This can be demonstrated with a simple example, finding the impedance of the circuit shown in Fig. 4.57.

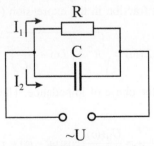

Figure 4.57. Impedance of parallel connection of the resistor and capacitor

First one needs to find the total complex current of the circuit $\hat{I} = \hat{I}_1 + \hat{I}_2$, where $\hat{I}_1 = \hat{U}/R$ and $\hat{I}_2 = dq/dt = d(C\hat{U})/dt = j\omega C\hat{U}$. According to the above definition of impedance

$$\hat{Z} = \frac{\hat{U}}{\hat{I}_1 + \hat{I}_2} = \frac{R}{1 + j\omega RC}. \tag{4.42}$$

But the same result can be obtained by calculating the impedance as the resistance of the parallel-connected conductors

$$\frac{1}{\widehat{Z}} = \frac{1}{\widehat{Z}_1} + \frac{1}{\widehat{Z}_2}, \text{ where } \widehat{Z}_1 = \frac{1}{j\omega C}, \ \widehat{Z}_2 = R. \tag{4.43}$$

Performing elementary transformations, one can obtain

$$\widehat{Z} = \frac{R}{1 + j\omega RC}. \tag{4.44}$$

This method of computation works for both parallel and serial connection of ideal elements, as well as their combinations. Its application allows calculating the impedance of complex electrochemical systems using simple calculation rules for parallel and series connection of conductors.

Impedance representation in phase plane

To describe dynamic systems, phase-plane portraits are widely used. The phase portrait of a dynamic system is the trajectory of a point depicting the state of the system on the phase plane or in the phase space at a certain moment. For phase trajectories plotting, generalized coordinates are used. For the simple pendulum Equation (4.23) as the generalized coordinates are spatial $x(t)$ coordinate and its velocity $\dot{x}(t)$. In the case of an RLC-circuit, as the generalized coordinates can be considered $q(t)$ and $\dot{q}(t)$, i.e., capacitor charge and electric current. Thus the phase plane for RLC-circuit is the Cartesian coordinate system q(t) and $\dot{q}(t)$. Comparing expressions (4.32) and (4.33), one can see that the generalized coordinates are related as $\dot{q}(\omega) = j\omega q(\omega)$ or $\dot{q}(\omega) = \omega q(\omega)e^{j\frac{\pi}{2}}$. In other words, the current modulus is ω times larger than the charge modulus and is ahead of the charge oscillations in phase by an angle $\pi/2$. It is known that if oscillations of the depicting point in the phase plane occur in two mutually perpendicular directions according to a harmonic law with different amplitudes and with a phase shift by $\pi/2$, then the trajectory of representation point is an ellipse. This is the phase portrait of an open system executing forced harmonic oscillations.

In the case of impedance, such a phase portrait does not make sense, since impedance is time independent. In this situation, it is more convenient to consider the frequency as an independent variable. If the impedance of the system can be represented in the form $\text{Re}(\widehat{Z}) = Z'(\omega)$, $\text{Im}(\widehat{Z}) = jZ''(\omega)$, where the real and imaginary parts of the impedance are known frequency functions, then such a system may be considered as a parametric equation of a phase trajectory in the complex plane Z',0,Z''. Such curves are referred to as Nyquist diagrams.

As an example, the already familiar electrical circuit consisting of capacitors and a resistor connected in parallel (Fig. 4.57) can be considered. Separating the real and imaginary parts in the expression for impedance (4.44), one can obtain the parametric equation of the curve known as the Nyquist diagram

$$\text{Re}\widehat{Z} = \frac{R}{1 + (\omega RC)^2}, \tag{4.45}$$

$$\text{Im}\widehat{Z} = -j\frac{\omega RC}{1 + (\omega RC)^2}. \tag{4.46}$$

The curve itself, having the form of an arc, is presented in Fig. 4.58. At zero frequency, as can be seen from the above expressions, $Re(\widehat{Z}) = R$, $Im(\widehat{Z}) = 0$. As the frequency increases, the representation point of the system goes to the upper half-plane (for convenience, the Nyquist diagrams are usually plotted as $-Im(\widehat{Z})$ versus $Re(\widehat{Z})$, outlines the arc and tends to the origin at $\omega \to \infty$.

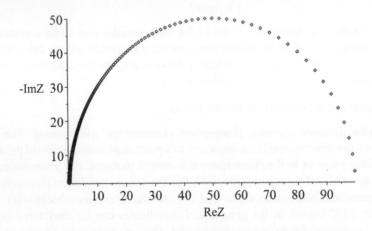

Figure 4.58. Nyquist diagram for RC circuit, shown in Fig. 4.57

Nyquist diagrams are plotted on the complex plane in the Cartesian coordinates, but complex numbers can also be represented in polar coordinates. In this case, the position of the representation point is determined by the modulus of the complex number $\rho(\omega) = \sqrt{Z'^2(\omega) + Z''^2(\omega)}$ and its phase $\varphi(\omega) = \arctan[Z''(\omega)/Z'(\omega)]$. A graphical representation of the impedance amplitude and phase dependences on the logarithm of the frequency is referred to as the Bode diagram. These diagrams are used along with the Nyquist diagrams and supplement them, as they reflect the dependence of the impedance on the frequency.

4.4.2 Electrochemical impedance

All the mathematical models of RLC-circuit discussed above are an example of the classical approach to the study of electrochemical systems [17]. In the framework of this approach, the objects under study are described by mathematical models in the form of systems of linear algebraic and differential equations, both ordinary and partial derivative equations. For their analysis, a wide range of methods, from analytical to numerical, are used. Although the classical approach is indispensable in conducting fundamental research, it is of little use for solving applied problems due to its low operativeness.

Another drawback of the classical approach is the impossibility of experimental verification of some models. The fact is that the theoretically obtained dependence of the impedance on the frequency often contains more parameters than can be determined from the analysis of experimental results. Thus, mathematical models are experimentally unverifiable. The way out of this situation is the structural approach, in which the electrochemical system is divided into separate components that interact with each other through surfaces and bonds that separate them. Each component of the system is associated with one of the elements of the electrical circuit that best reflects the properties of this component. These can be both the electrotechnical elements discussed above and elements not used in electrical engineering. If the result is an electrical circuit consisting of idealized circuit elements for which the calculated voltages and currents coincide with the currents and voltages of real elements with a given accuracy, then such a circuit is called an equivalent circuit of a real electrochemical system.

Within the framework of the structural approach, the main task is to select an equivalent circuit that could contain the minimum number of elements and best describe the experimental data. Its mathematical model will contain exactly as many parameters as can be determined proceeding from experimental data. This allows one to calculate the frequency dependences of the impedance and compare them with the experimental dependences. In this sense, structural models are flexible and operational, and even relatively simple impedance spectrometers can generate equivalent circuits that best approximate experimental diagrams.

Below are considered mathematical models of some idealized elements of which equivalent circuits of real electrochemical systems can consist. Examples of calculating the impedance of these elements, the corresponding Nyquist diagrams and their interpretation are given.

Ideally polarizable electrode

An electrode through the surface of which there is no transfer of electric charge is called an ideally polarizable or blocking electrode. If such an electrode is immersed in a solution of a supporting electrolyte, then when a potential is supplied to it, an electric current will flow through the electrolyte, forming a double electric layer. Accordingly, the equivalent circuit of such

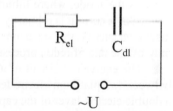

Figure 4.59. Equivalent circuit of an ideally polarizable electrode

a system will consist of the resistance of an electrolyte connected in series with a capacitor simulating the capacitance of a double electric layer (Fig. 4.59).

The impedance of such systems is calculated as the sum of the corresponding complex resistances and is described by the above expression (4.38). In this expression, the real and imaginary parts are already separated, so the form of the Nyquist and Bode diagrams shown in Fig. 4.60 is quite obvious.

Electrode with Faraday process

An example of an ideal polarizable electrode is platinum electrodes in aqueous solutions, which thermodynamically cannot enter in electrochemical reactions in a certain range of potentials. An example is an aqueous solution of NaF in the potential range from 0 to 1.2 V relative to a normal hydrogen electrode. Platinum electrodes are widely used to measure the conductivity of solid electrolytes, such as LiPON. According to the data of [18], in this case they are not ideal, because at a potential of

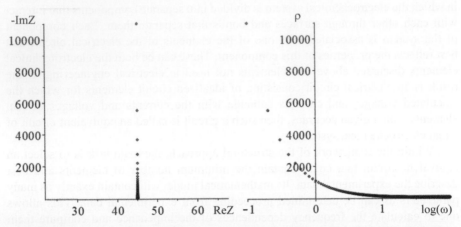

Figure 4.60. Nyquist (left) and Bode (right) diagrams of an ideally polarizable electrode calculated for $R_{el} = 45\ \Omega$ and $C = 0.9$ mF

1 V, the leakage current density is 0.91×10^{-6} A/cm² while corresponding to leakage resistance is $R_{lk} = 1.76 \times 10^6\ \Omega$. Thus, the Faraday process takes place in the system, which occurs as follows. Under the action of an electric field, lithium ions move to the cathode, where they are reduced, increasing the concentration of neutral lithium atoms. In accordance with Fick's first law, this results in a diffusion flow of lithium towards the anode, where lithium is oxidized and the process is repeated. Thus the low leakage current is primarily due to the resistance of diffusion flow, which will be discussed later. At this stage, one can assume that the leakage resistance is determined only by the rate of redox processes on the electrodes.

The equivalent circuit of a solid electrolyte with ohmic resistance R_{el}, located between two platinum electrodes, on the surface of which, with voltage application, a double electric layer of the capacitance C_{dl} is formed can be considered (Fig. 4.59). The Faraday process in the circuit is modeled by leakage resistance R_{lk}.

The impedance of the circuit shown in Fig. 4.61 is calculated as the total resistance of parallel and series-connected conductors

$$\hat{Z} = R_{el} + \hat{Z}_1, \text{ where } \frac{1}{\hat{Z}_1} = \frac{1}{R_{lk}} + \frac{1}{\hat{Z}_C}, \ \hat{Z}_C(\omega) = -j\frac{2}{\omega C_{dl}}. \tag{4.47}$$

Performing of elementary reduction in (4.47) yields

$$\hat{Z} = \frac{R_{el} + R_{lk} + j\omega R_{el} R_{lk} C_{dl} / 2}{1 + j\omega R_{lk} C_{dl} / 2}. \tag{4.48}$$

Separating of the real and imaginary parts in (4.48), results in the parametric equation of the Nyquist diagram

$$\operatorname{Re}\hat{Z} = \frac{R_{lk} + R_{el}(1 + \omega^2 R_{lk}^2 C_{dl}^2 / 4)}{1 + \omega^2 R_{lk}^2 C_{dl}^2 / 4}, \tag{4.49}$$

$$\operatorname{Im}\hat{Z} = -j\frac{\omega R_{lk} C_{dl} / 2}{1 + (\omega R_{lk} C_{dl} / 2)^2} R_{lk}. \tag{4.50}$$

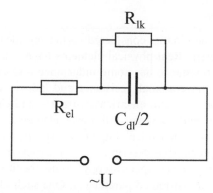

Figure 4.61. Equivalent circuit of electrolyte thin film with two electrodes regarding the Faraday process

The Nyquist diagram, calculated for the experimental values of the circuit parameters $R_{el}= 50\ \Omega$, $R_{lk}= 1.1\cdot10^4\Omega$, $C_{dl}= 2.0\cdot10^{-7}F$, is presented in Fig. 4.62.

The limiting case of the considered system is an ideal non-polarizable electrode. This occurs when the leakage resistance is small compared to the resistance of the electrolyte. The impedance of the system is in this case can be found. Calculating the limits of expressions (4.49) and (4.50) with the leakage resistance tending to zero, one can obtain $\operatorname*{Re}_{R_{lk}\to 0}\hat{Z} \to R_{el}$ and $\operatorname*{Im}_{R_{lk}\to 0}\hat{Z} \to 0$. Thus if the leakage resistance is sufficiently small, then the impedance of the system is real and equal to the resistance of the electrolyte.

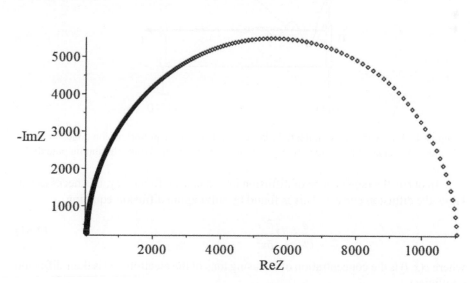

Figure 4.62. Nyquist diagram for a system with a Faraday process (Fig. 4.61) in the frequency range from 10 Hz to 20 kHz. The curve leaves the real axis at the point $\operatorname{Re}\hat{Z}_1 = R_{el} + R_{lk}$ and indicates to $\operatorname{Re}\hat{Z}_2 = R_{el}$ at $\omega \to \infty$

Warburg impedance

The elements of all the circuits considered earlier are ideal in the sense that they have lumped parameters. Real physical elements have distributed parameters, and for this reason, real resistance has some inductance and capacitance. In the same way, capacitor plates have a small resistance and inductance. The same applies to inductance, which has a certain electrical resistance and capacitance, since it may accumulate excess electric charge. All these elements, called electrotechnical, are widely used in electrochemical impedance spectroscopy. However, they are insufficient for comprehensive description of electrochemical systems. Therefore, in electrochemical impedance spectroscopy, some specific elements are used for modeling systems with distributed parameters. One such element is a semi-infinite diffusion element, referred to as Warburg diffusion element [19].

The electrochemical processes that Warburg impedance models can be briefly considered. A sinusoidal voltage is applied to one of the electrodes of the electrochemical cell as shown in Fig. 4.63. At one polarity of the voltage, the electrode dissolves with the formation of excess ions, and at the other polarity, the ions are reduced and deposited on the electrode. In the considered model, it is assumed that the macroscopic field in the electrolyte is negligibly small due to its high conductivity. As a result ions drift through the electrolyte only due to diffusion.

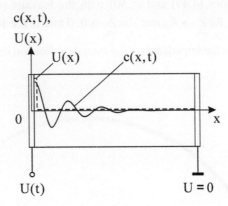

Figure 4.63. Distribution of potential (dashed line) and non equilibrium ion concentration (solid curve) in an electrolyte with high conductivity when applying a harmonic potential

To obtain the dependence of diffusion impedance on frequency, it is necessary to know the diffusion current. This is found by solving the diffusion equation

$$\frac{\partial c}{\partial t} = D\frac{\partial^2 c}{\partial x^2}, \tag{4.51}$$

where $c(x, t)$ is the concentration of diffusing ions of the electrode, D is their diffusion coefficient.

Within the model, the input voltage changes the electrode potential according to the harmonic law $U(t) = U_0 e^{j\omega t}$, where U_0 is the potential amplitude. In addition, it is assumed that excess ions are formed only on the electrode (in plane $x = 0$), and their concentration in this plane is proportional to the applied voltage and changes in phase with the voltage. Hence, the first boundary condition should look as follows

$$c(0,t) = c_0 e^{j\omega t} \tag{4.52}$$

where c_0 is a real number, which follows from the assumption of a co-phased oscillation of voltage and concentration. For a semi-space, the second boundary condition has the form

$$\lim_{x \to \infty} c(x,t) = 0. \tag{4.53}$$

If a diffusion region is limited then as the second boundary condition, two cases are usually considered. The first is the impermeability condition, which is also called the reflective boundary condition, because it means that at the boundary, the forward and reverse diffusion fluxes are equal, and the concentration gradient is zero

$$\left. \frac{\partial c}{\partial x} \right|_{x=\delta} = 0. \tag{4.54}$$

Such a condition can be set for a system in which the second electrode in a plane $x = \delta$ (Fig. 4.63) is a blocking. Then the concentration of nonequilibrium ions on the electrode will be greater than zero, while its gradient will be zero.

The second condition is a transparency condition or a zero-boundary condition, which requires the nonequilibrium concentration of ions at the second boundary, is zero

$$c(\delta,t) = 0. \tag{4.55}$$

Such a condition is applicable, for instance, to a system in which the Faraday process takes place.

Warburg Impedance for semi-infinite diffusion

First one needs to consider the case of a semi-infinite medium, i.e., boundary value problem (4.51)-(4.53). The search for the settled periodic solution is carried out by the method of separation of variables, according which the desired solution is presented in a form of a product

$$c(x,t) = A(x)e^{\mu t}, \tag{4.56}$$

where $\mu = j\omega$, that corresponds to the steady-state oscillation mode. Substituting (4.56) into (4.51), one can obtain the following equation

$$DA''(x) - \mu A(x) = 0. \tag{4.57}$$

Its particular solution is tried in the form $A = \exp(\lambda x)$, whose substitution in (4.57) reveals that the parameter λ (of the dimension of inverse length) has two values.

$$\lambda = \pm \sqrt{\frac{\mu}{D}}. \tag{4.58}$$

The general solution of Equation (4.57) has the form of a linear combination of particular solutions

$$A(x) = C_1 e^{\sqrt{\frac{\mu}{D}}x} + C_2 e^{-\sqrt{\frac{\mu}{D}}x}. \tag{4.59}$$

If the right boundary is so far from the electrode that the ion density waves decay before reaching it, then one can use the second boundary condition (4.53). From relation (4.59) it follows that as $x \to \infty$, the first term grows unboundedly, while the second term exponentially decays. Therefore, the second boundary condition holds only if $C_1 = 0$. Then solution (4.56) takes the form

$$c(x,t) = C_2 e^{-(1+j)\beta x} e^{j\omega t}, \tag{4.60}$$

where $\beta = \sqrt{\omega/2D}$. The second constant of integration, derived from the first boundary condition, is $C_2 = c_0$, that allows one to write down the final form of the solution of the boundary value problem (4.51)-(4.53)

$$c(x,t) = c_0 e^{-(1+j)\beta x} e^{j\omega t}. \tag{4.61}$$

The next step is to express the diffusion current density from Fick's first law $\bar{j} = -D\partial c / \partial x$ as

$$\bar{j} = (1+j)\beta c_0 D e^{-(1+j)\beta x} e^{j\omega t} \tag{4.62}$$

and find the electric current

$$I(x, t) = (1+j)SqD\beta c_0 e^{-(1+j)\beta x} e^{j\omega t}, \tag{4.63}$$

where S is the electrode area, q is the ion charge. By definition, the expression for the impedance should be

$$Z(x) = \frac{U(t)}{I(x,t)} = \frac{U_0 e^{(1+j)\beta x}}{(1+j)SqD\beta c_0}. \tag{4.64}$$

Since within the framework of the approach used, the parameters of the circuit are assumed to be concentrated, in the expression (4.64) one could assume $x = 0$. In this case, expression (4.64) takes the form

$$Z(0) = \frac{U_0}{(1+j)SqD\beta c_0} = \frac{\sqrt{2}}{1+j} \frac{U_0}{\sqrt{\omega D} Sqc_0} = (1-j)\frac{U_0}{\sqrt{2\omega D} Sqc_0}. \tag{4.65}$$

In the literature, it is usually given as

$$\widehat{Z}_W = \frac{U_0}{\sqrt{2\omega D} Sqc_0}(1-j) = \frac{A_W}{\sqrt{\omega}} - j\frac{A_W}{\sqrt{\omega}} = \frac{A_W}{\sqrt{\omega}} + \frac{A_W}{j\sqrt{\omega}} \tag{4.66}$$

and is referred to as Warburg impedance, and the parameter $A_W = \dfrac{U_0}{\sqrt{2D} Sqc_0}$ is called Warburg coefficient.

Note that the phase of the Warburg element $\varphi = \arctan(\mathrm{Im}Z_W / \mathrm{Re}Z_W) = \arctan(-1)$ is constant and equal $-\pi/4$. Therefore, all Nyquist diagrams for the Warburg element are independent of the Warburg coefficient and have the form of a straight line inclined at an angle of 45°. This is illustrated in the diagram in Fig. 4.64, while Fig 4.65 shows the dependence of the Warburg impedance modulus on the logarithm of the angular frequency (Bode diagram).

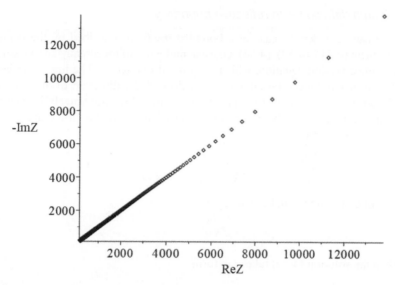

Figure 4.64. Nyquist diagram for a Warburg element in the frequency range 1 Hz-1 kHz

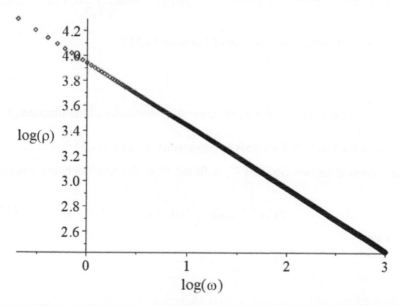

Figure 4.65. Bode diagram showing the change in the decimal logarithm of the modulus of the Warburg element with coefficient $A_W = 6176$ $\Omega \cdot c^{-1/2}$ in the frequency range 0.1 Hz-1 kHz

It must be emphasized once again that the Warburg element is an element of the structural model that describes some properties of a real system but have a limited scope. According to Fig. 4.64, $\mathrm{Im}\,\widehat{Z} \to \infty$ as $\omega \to 0$ while for any real electrochemical cell, the imaginary part of the impedance should tend to zero, as resistance to direct current is a real value.

Finite-length diffusion with reflective boundary

Now one can consider the case of a bounded medium described by the boundary value problem (4.51), (4.52), (4.54). One can make use of the already known solution (4.59) of the diffusion equation, which consists of two terms. The first describes the distribution of ion concentration, creating a flow in the direction of the x axis, the second describes the flow in the opposite direction. The impermeability condition (4.54) means that at the boundary the forward and reverse flows are equal in magnitude. The condition

$$\frac{\partial A(x)}{\partial x}\bigg|_{x=\delta} \equiv C_1 \sqrt{\frac{\mu}{D}} e^{\sqrt{\frac{\mu}{D}}\delta} - C_2 \sqrt{\frac{\mu}{D}} e^{-\sqrt{\frac{\mu}{D}}\delta} = 0 \qquad (4.67)$$

can be applied to find the unknown constant

$$C_2 = C_1 e^{2\sqrt{\frac{\mu}{D}}\delta}. \qquad (4.68)$$

Then the solution (4.59) takes the form

$$A(x) = C_1 e^{\sqrt{\frac{\mu}{D}}\delta} \left(e^{\sqrt{\frac{\mu}{D}}(\delta-x)} + e^{-\sqrt{\frac{\mu}{D}}(\delta-x)} \right) = 2C_1 e^{\sqrt{\frac{\mu}{D}}\delta} \cosh\left(\sqrt{\frac{\mu}{D}}(\delta-x)\right) \qquad (4.69)$$

which allows one to obtain the solution of Equation (4.51)

$$c(x,t) = 2C_1 e^{\sqrt{\frac{\mu}{D}}\delta} \cosh\left(\sqrt{\frac{\mu}{D}}(\delta-x)\right) e^{\mu t}. \qquad (4.70)$$

Note that in the plane $x = \delta$ only the current or derivative of the function (4.70) vanishes, then the function itself does not disappear $c(\delta,t) = 2C_1 e^{\sqrt{\frac{\mu}{D}}\delta} e^{\mu t}$.

The second unknown constant C_1 is found from the first boundary condition (4.52)

$$2C_1 e^{\sqrt{\frac{\mu}{D}}\delta} \cosh\left(\sqrt{\frac{\mu}{D}}\delta\right) = c_0, \qquad (4.71)$$

which yields

$$2C_1 = \frac{c_0}{e^{\sqrt{\frac{\mu}{D}}\delta} \cosh\left(\sqrt{\frac{\mu}{D}}\delta\right)}. \qquad (4.72)$$

Then the solution of Equation (4.51) takes the final form

$$c(x,t) = c_0 \frac{\cosh\left(\sqrt{\frac{\mu}{D}}(\delta-x)\right)}{\cosh\left(\sqrt{\frac{\mu}{D}}\delta\right)} e^{\mu t}. \qquad (4.73)$$

It is easy to see that it meets the boundary condition (4.52).

From the solution obtained and the first Fick law the diffusion current density

$$\bar{j} = Dc_0 \sqrt{\frac{\mu}{D}} \frac{\sinh\left(\sqrt{\frac{\mu}{D}}(\delta - x)\right)}{\cosh\left(\sqrt{\frac{\mu}{D}}\delta\right)} e^{\mu t} \tag{4.74}$$

and the total electric current

$$I(x,t) = qSDc_0 \sqrt{\frac{\mu}{D}} \frac{\sinh\left(\sqrt{\frac{\mu}{D}}(\delta - x)\right)}{\cosh\left(\sqrt{\frac{\mu}{D}}\delta\right)} e^{\mu t} \tag{4.75}$$

can be obtained. One could see that the solution really meets the impermeability condition. Indeed with $x = \delta$ a hyperbolic sine, vanishes and therefore, the current at the boundary will be zero.

At the plane $x = 0$, the current will be equal to

$$I(0,t) = qSDc_0 \sqrt{\frac{\mu}{D}} \tanh\left(\sqrt{\frac{\mu}{D}}\delta\right) e^{\mu t}. \tag{4.76}$$

From the resulting ratio, follows the expression for Warburg impedance

$$\hat{Z} = \frac{U(0,t)}{I(0,t)} = (1-j) \frac{U_0}{\sqrt{2\omega D} qSc_0} \frac{\sinh\left(\sqrt{\frac{2\omega}{D}}\delta\right) - j\sin\left(\sqrt{\frac{2\omega}{D}}\delta\right)}{\cosh\left(\sqrt{\frac{2\omega}{D}}\delta\right) - \cos\left(\sqrt{\frac{2\omega}{D}}\delta\right)}. \tag{4.77}$$

Separation of the real and imaginary parts of Equation (4.77) yields

$$\operatorname{Re}(\hat{Z}) = \frac{A_W}{\sqrt{\omega}} \frac{\sinh\left(\sqrt{\frac{2\omega}{D}}\delta\right) - \sin\left(\sqrt{\frac{2\omega}{D}}\delta\right)}{\cosh\left(\sqrt{\frac{2\omega}{D}}\delta\right) - \cos\left(\sqrt{\frac{2\omega}{D}}\delta\right)}, \tag{4.78}$$

$$\operatorname{Im}(\hat{Z}) = -j\frac{A_W}{\sqrt{\omega}} \frac{\sinh\left(\sqrt{\frac{2\omega}{D}}\delta\right) + \sin\left(\sqrt{\frac{2\omega}{D}}\delta\right)}{\cosh\left(\sqrt{\frac{2\omega}{D}}\delta\right) - \cos\left(\sqrt{\frac{2\omega}{D}}\delta\right)}. \tag{4.79}$$

In cases of thick electrolyte layer, low diffusion coefficient or high frequency, i.e., whenever the condition $\delta \gg \sqrt{D/\omega}$ is met, the oscillations in the ions density decay very quickly and cease affecting the concentration of ions in a vicinity of the boundary $x = \delta$. That is, the medium in the framework of the boundary value problem becomes semi-infinite and the final-length diffusion impedance tends to the Warburg impedance. The hyperbolic sine and cosine functions are unbounded, and at high frequencies the sine and cosine in expression (4.77) can be neglected. Since the limit of the ratio of the hyperbolic sine to the hyperbolic cosine when the argument tends to infinity is equal to unity, the limit of expression (4.77) takes the form

$$\lim_{\omega \to \infty} \hat{Z}(\omega) = (1 - j)\frac{U_0}{\sqrt{2\omega D q S}\, A_0} = \hat{Z}_W. \tag{4.80}$$

Figure 4.66 shows the Nyquist diagrams for a finite diffusion element with the condition of impermeability at the boundary. At low frequencies (Fig. 4.66a), the influence of the boundary leads to a deviation from the Warburg impedance. In a high frequency range (Fig. 4.66b), the impedance of the finite diffusion element tends to Warburg impedance. The calculations were performed for an element consisting of a solid electrolyte of a thickness $\delta = 1\mu m$, a diffusion coefficient $D \sim 5 \times 10^{-10}$ cm²·s and Warburg coefficient $A_W = 6176$ Ω·s⁻¹/².

Finite-length diffusion element with transmissive boundary

One can consider the case when the right boundary $x = \delta$ is transparent to the diffusion flow. This is possible in the system where the Faraday process takes place, or the corresponding electrode has a sufficiently large capacitance and diffusion conductivity. In this case, the non equilibrium ion concentration vanishes at $x = \delta$, while its derivative is not zero $\partial A / \partial x\big|_{x=\delta} \neq 0$. To determine the integration constants in solution (4.59), it is necessary to use the zero-boundary condition (4.55)

$$A(\delta) = C_1 e^{\sqrt{\frac{\mu}{D}}\delta} + C_2 e^{-\sqrt{\frac{\mu}{D}}\delta} = 0 \tag{4.81}$$

from which it follows that

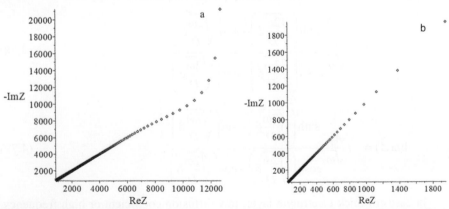

Figure 4.66. Nyquist diagram for the finite-length diffusion element with reflective boundary: a - low frequencies (0.07 Hz-10 Hz), b - medium and high frequencies (10 Hz-1 kHz)

$$C_2 = -C_1 e^{2\sqrt{\frac{\mu}{D}}\delta}. \tag{4.82}$$

Then the spatial part of the solution takes the form

$$A(x) = -2C_1 e^{\sqrt{\frac{\mu}{D}}\delta} \sinh\left(\sqrt{\frac{\mu}{D}}(\delta - x)\right). \tag{4.83}$$

The constant $2C_1$ determined from the second boundary condition

$$-2C_1 e^{\sqrt{\frac{\mu}{D}}\delta} \sinh\left(\sqrt{\frac{\mu}{D}}\delta\right) = c_0 \tag{4.84}$$

as

$$-2C_1 = \frac{c_0}{e^{\sqrt{\frac{\mu}{D}}\delta} \sinh\left(\sqrt{\frac{\mu}{D}}\delta\right)}, \tag{4.85}$$

allows one to obtain the final form of the solution of the boundary value problem (4.51), (4.52), (4.55)

$$c(x,t) = c_0 \frac{\sinh\left(\sqrt{\frac{\mu}{D}}(\delta - x)\right)}{\sinh\left(\sqrt{\frac{\mu}{D}}\delta\right)} e^{\mu t}. \tag{4.86}$$

From the obtained expression, it is possible to find the density of diffusion flow and, multiplying it by the ion charge and electrode area, find the current through the electrolyte

$$I(0,t) = qSDc_0 \sqrt{\frac{\mu}{D}} \coth\left(\sqrt{\frac{\mu}{D}}\delta\right) e^{\mu t}. \tag{4.87}$$

It remains only to divide the voltage across the electrode by the current to obtain an expression for the finite-length diffusion impedance in a system with a transmissive boundary

$$\hat{Z} = (1-j)\sqrt{\frac{D}{2\omega}} \frac{U_0}{qSDc_0} \frac{\sinh\left(\sqrt{\frac{2\omega}{D}}\delta\right) + j\sin\left(\sqrt{\frac{2\omega}{D}}\delta\right)}{\cosh\left(\sqrt{\frac{2\omega}{D}}\delta\right) + \cos\left(\sqrt{\frac{2\omega}{D}}\delta\right)}. \tag{4.88}$$

After separating the variables, one gets

$$\mathrm{Re}(\hat{Z}) = \frac{A_W}{\sqrt{\omega}} \frac{\sinh\left(\sqrt{\frac{2\omega}{D}}\delta\right) + \sin\left(\sqrt{\frac{2\omega}{D}}\delta\right)}{\cosh\left(\sqrt{\frac{2\omega}{D}}\delta\right) + \cos\left(\sqrt{\frac{2\omega}{D}}\delta\right)}, \tag{4.89}$$

$$\text{Im}(\widehat{Z}) = -j\frac{A_W}{\sqrt{\omega}}\frac{\sinh\left(\sqrt{\frac{2\omega}{D}}\delta\right) - \sin\left(\sqrt{\frac{2\omega}{D}}\delta\right)}{\cosh\left(\sqrt{\frac{2\omega}{D}}\delta\right) + \cos\left(\sqrt{\frac{2\omega}{D}}\delta\right)}. \tag{4.90}$$

When $\omega \to \infty$ the impedance of finite-length diffusion element, as in the previous case, tends to the Warburg impedance $\widehat{Z}_W = (1-j)\dfrac{U_0}{\sqrt{2\omega DqSc_0}}$.

Figure 4.67 shows the Nyquist diagrams for a finite-length diffusion element with a transmissive boundary. At low frequencies (Fig. 4.67a), the influence of the boundary results in a deviation from the Warburg impedance. In the high frequency range (Fig. 4.67b) the impedance tends to that of Warburg. The calculations were performed for an element consisting of a solid electrolyte with a thickness $\delta = 1$ μm, a diffusion coefficient $D \sim 5\times10^{-10}$ cm$^2 \cdot$ s^{-1} and Warburg coefficient $A_W = 6.176\times10^3$ Ω c$^{-1/2}$.

Gerischer impedance

The Warburg impedance models processes in the electrolyte, in which the number of ions in the electrolyte changes only in the near-electrode region due to its dissolution and deposition or due to the oxidation and reduction of ions. In a solid electrolyte, the local concentration of ions can also change due to the activation of ions or their capture by vacancies. These are processes of generation and recombination of charge carriers which are considered within the Gerischer model. A brief history of the term 'Gerischer impedance' is given in [20]. In 1951, Gerischer published the work [21], which gave a formal interpretation of the response of direct and alternating current to a sequential chemical and electrochemical reaction on an inert electrode in an aqueous electrolyte. Almost 20 years later, Sluyters-Rehbach and Sluyters [22] presented a general expression of the impedance for such a reaction. In 1984, they published an expression of the impedance for a reaction of the CEC type (Chemical-Electrochemical-Chemical) [23], which was called the Gerischer impedance. Under

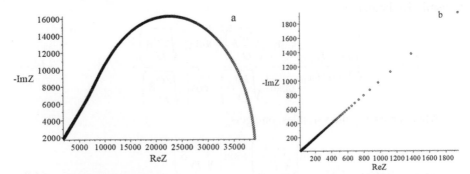

Figure 4.67. Nyquist diagram for the finite-length diffusion element with transmissive boundary: a - low frequencies (7 mHz-10 Hz), b - medium frequencies (10 Hz-10 kHz)

special conditions [23], this complex expression reduces to a rather simple impedance relation:

$$Z(\omega) = \frac{Z_0}{\sqrt{k + j\omega}} \, . \tag{4.91}$$

According to [20], the diffusion model of Gerischer differs from the Warburg model by the additional term $- kc(x, t)$. This term considers the rate of chemical reactions, resulting in loss of ions' mobility (Fig. 4.68). Such terms of continuity equations, to which the diffusion equation belongs, are referred to as volumetric sources or sinks, depending on the sign. In our case, $- kc(x, t)$ is the volumetric density of the sinks. In the relaxation time approximation, this density can be expressed as $- c(x, t)/\tau$, where τ is the relaxation time of the nonequilibrium concentration due to chemical reactions or recombination of charge carriers. This approximation is rather rough and is performed more accurately, the smaller the deviation $c(x, t)$ from the equilibrium concentration value.

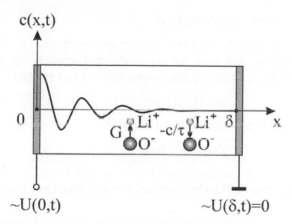

Figure 4.68. Forced oscillations in the concentration of lithium in a solid electrolyte and chemical reactions in the volume of the electrolyte

The diffusion equation, or Fick's second law, belongs to balance equations and, to consider both oxidation and reduction processes, must also contain the function of sources with a '+' sign. In linear approximation, this is simply a constant that in the course of calculations is included in the stationary (equilibrium) concentration $c(x)$ and does not appear in the equation for the non equilibrium concentration. Therefore, for a nonequilibrium process, Equation (4.51) takes the form

$$\frac{\partial c}{\partial t} = D \frac{\partial^2 c}{\partial x^2} - \frac{c}{\tau}, \tag{4.92}$$

The boundary conditions in this boundary value problem remain the same as in problem (4.51)-(4.55). So, the first boundary condition has the form

$$c(0, t) = c_0 e^{j\omega t}, \tag{4.93}$$

and the second boundary condition for semi-space is the condition of concentration waves decay

$$\lim_{x \to \infty} c(x,t) = 0.$$ (4.94)

For a finite-length diffusion region at the second boundary in the case of a blocking electrode, the impermeability condition must be satisfied

$$\left. \frac{\partial c}{\partial x} \right|_{x=\delta} = 0.$$ (4.95)

The second limiting case is when the electrolyte is in contact with an infinitely large reservoir. Then the boundary condition for the non equilibrium concentration takes the form

$$c(\delta, t) = 0.$$ (4.96)

Gerischer impedance for semi-space

One could first consider the case of a semi-infinite medium, i.e., boundary value problem (4.92)-(4.94) and look for the established solution by the method of separation of variables, presenting the desired solution as a product

$$c(x, t) = A(x)e^{j\omega t}.$$ (4.97)

Substitution of (4.97) in (4.92) yields the equation

$$DA''(x) - \gamma A(x) = 0,$$ (4.98)

where $\gamma = 1/\tau + j\omega$. Its particular solution is sought for as $A = \exp(\lambda x)$ and is substituted in Equation (4.98) which reduces it to a characteristic equation. The latter has two roots

$$\lambda_{1,2} = \pm \sqrt{\frac{\gamma}{D}}.$$ (4.99)

The general solution of Equation (4.98) has the form of a linear combination of particular solutions

$$A(x) = C_1 e^{\sqrt{\frac{\gamma}{D}}x} + C_2 e^{-\sqrt{\frac{\gamma}{D}}x}.$$ (4.100)

According to the second boundary condition, the solution (4.100) should decay at infinity, which is possible only under the condition of $C_1 = 0$. Then the solution (4.97) takes the form

$$c(x,t) = C_2 e^{-\sqrt{\frac{1}{\tau D} + j\frac{\omega}{D}}x} e^{j\omega t}.$$ (4.101)

The second constant of integration is defined from the first boundary condition as $C_2 = c_0$, which allows one to write down the final form of the solution of the boundary value problem (4.92)-(4.94)

$$c(x,t) = c_0 e^{-\sqrt{\frac{1}{\tau D}+j\frac{\omega}{D}}x} e^{j\omega t}. \tag{4.102}$$

The diffusion current density, expressed from the first Fick's law

$$\bar{j} = c_0 \sqrt{\frac{D}{\tau} + j\omega D} \, e^{-\sqrt{\frac{1}{\tau D}+j\frac{\omega}{D}}x} e^{j\omega t} \tag{4.103}$$

is easily converted to the electric current

$$I(x,t) = Sqc_0 \sqrt{\frac{D}{\tau} + j\omega D} \, e^{-\sqrt{\frac{1}{\tau D}+j\frac{\omega}{D}}x} e^{j\omega t}, \tag{4.104}$$

where S is electrode area, q is the ion charge.

In accordance with the definition of impedance, as the ratio of voltage to current, one can obtain the following expression

$$\widehat{Z}_G(x) = \frac{U(t)}{I(x,t)} = \frac{U_0}{Sqc_0\sqrt{D}} \frac{e^{\sqrt{\frac{1}{\tau D}+j\frac{\omega}{D}}x}}{\sqrt{\frac{1}{\tau} + j\omega}}. \tag{4.105}$$

At last, equating $x = 0$ one can obtain the final form of the Gerischer impedance

$$\widehat{Z}_G = \frac{U_0}{Sqc_0\sqrt{D}} \frac{1}{\sqrt{\frac{1}{\tau} + j\omega}}. \tag{4.106}$$

To separate the real and imaginary parts, the square root formula of the complex number is required

$$\sqrt{a+jb} = \frac{1}{\sqrt{2}}\sqrt{\rho+a} + j\frac{1}{\sqrt{2}}\sqrt{\rho-a}, \tag{4.107}$$

where $\rho = \sqrt{a^2 + b^2}$. Its application to Equation (4.106) yields

$$Z_G = \frac{A_W}{\rho}\left(\sqrt{\rho+1/\tau} - j\sqrt{\rho-1/\tau}\right), \tag{4.108}$$

where $\rho = \sqrt{1/\tau^2 + \omega^2}$, $A_W = \dfrac{U_0}{\sqrt{2D}Sqc_0}$ is the Warburg coefficient.

From the obtained expression it can be seen that the limit of the ratio of the imaginary and real parts of the impedance, like that of the Warburg element, is

$$\lim_{\omega\to\infty}\frac{\mathrm{Im}\,Z_G}{\mathrm{Re}\,Z_G} = -\lim_{\omega\to\infty}\sqrt{\frac{\rho(\omega)-1/\tau}{\rho(\omega)+1/\tau}} = -1. \tag{4.109}$$

The obtained results are illustrated by the Nyquist diagram for the Gerischer impedance in the frequency range from 0.1 Hz to 5 kHz (Fig. 4.69).

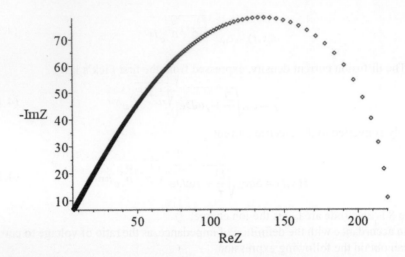

Figure 4.69. Nyquist diagram for the Gerischer impedance in the frequency range
1 Hz-5 kHz with system parameters $A_W = 1562.5.s^{-1/2}$, $\tau = 0.01s$

Gerischer impedance for a finite-length diffusion element with the impermeable boundary

Now one can consider the case of a bounded medium described by the boundary value problem (4.92), (4.93), (4.95). Since all subsequent arguments and calculations repeat the algorithm for solving the boundary value problem (4.51), (4.52), (4.54), starting from Equations (4.67) and till (4.76), only the calculations are given below.

One can use the already known solution (4.100) of the diffusion Equation (4.92) with the boundary condition (4.95)

$$\frac{\partial A(x)}{\partial x}\bigg|_{x=\delta} \equiv C_1\sqrt{\frac{\gamma}{D}}e^{\sqrt{\frac{\gamma}{D}}\delta} - C_2\sqrt{\frac{\gamma}{D}}e^{-\sqrt{\frac{\gamma}{D}}\delta} = 0, \tag{4.110}$$

and find the constant

$$C_2 = C_1 e^{2\sqrt{\frac{\gamma}{D}}\delta}. \tag{4.111}$$

Then the solution of Equation (4.92) transforms in the

$$c(x,t) = 2C_1 e^{\sqrt{\frac{\gamma}{D}}\delta}\cosh\left(\sqrt{\frac{\gamma}{D}}(\delta - x)\right)e^{j\omega t}. \tag{4.112}$$

The unknown constant C_1 can be expressed from the boundary condition (4.93) as

$$2C_1 = \frac{c_0 e^{-\sqrt{\frac{\gamma}{D}}\delta}}{\cosh\left(\sqrt{\frac{\gamma}{D}}\delta\right)}, \tag{4.113}$$

that allows to obtain the final solution of the boundary value problem (4.92), (4.93), (4.95).

$$c(x,t) = c_0 \frac{\cosh\left(\sqrt{\frac{\gamma}{D}}(\delta - x)\right)}{\cosh\left(\sqrt{\frac{\gamma}{D}}\delta\right)} e^{j\omega t}. \tag{4.114}$$

By differentiating the solution (4.114) with respect to x, one can find the ion flux density at the point $x = 0$, and through it express the electric current

$$I(0,t) = qSDc_0 \sqrt{\frac{\gamma}{D}} \tanh\left(\sqrt{\frac{\gamma}{D}}\delta\right) e^{j\omega t}. \tag{4.115}$$

By dividing the complex voltage by the complex current, one can find the Gerischer impedance for the finite-length diffusion element for the case of the impermeable boundary

$$\hat{Z}(\omega) = \frac{U_0}{qS\sqrt{Dc_0}\sqrt{1/\tau + j\omega}} \coth\left(\frac{\delta}{\sqrt{D}}\sqrt{1/\tau + j\omega}\right). \tag{4.116}$$

To separate the real and imaginary parts of the expression (4.116), one can take a square root of a complex number using the above formula (4.107). Its substitution into Equation (4.116) results in the expression

$$\hat{Z}(\omega) = \frac{U_0}{\sqrt{2D}Sqc_0\rho} \left(\sqrt{\rho + \frac{1}{\tau}} - j\sqrt{\rho - \frac{1}{\tau}}\right) \frac{\sinh\left(2\frac{\delta\sqrt{\rho + 1/\tau}}{\sqrt{2D}}\right) - j\sin\left(2\frac{\delta\sqrt{\rho - 1/\tau}}{\sqrt{2D}}\right)}{\cosh\left(2\frac{\delta\sqrt{\rho + 1/\tau}}{\sqrt{2D}}\right) - \cos\left(2\frac{\delta\sqrt{\rho - 1/\tau}}{\sqrt{2D}}\right)}, \tag{4.117}$$

which is easily divided into two parts: real

$$\operatorname{Re}\hat{Z}(\omega) = \frac{A_W}{\rho} \frac{\sqrt{\rho + 1/\tau}\sinh\left(2\frac{\delta\sqrt{\rho + 1/\tau}}{\sqrt{2D}}\right) - \sqrt{\rho - 1/\tau}\sin\left(2\frac{\delta\sqrt{\rho - 1/\tau}}{\sqrt{2D}}\right)}{\cosh\left(2\frac{\delta\sqrt{\rho + 1/\tau}}{\sqrt{2D}}\right) - \cos\left(2\frac{\delta\sqrt{\rho - 1/\tau}}{\sqrt{2D}}\right)} \tag{4.118}$$

and imaginary

$$\operatorname{Im}\hat{Z}(\omega) = -\frac{A_W}{\rho} \frac{\sqrt{\rho - 1/\tau}\sinh\left(2\frac{\delta\sqrt{\rho + 1/\tau}}{\sqrt{2D}}\right) + \sqrt{\rho + 1/\tau}\sin\left(2\frac{\delta\sqrt{\rho - 1/\tau}}{\sqrt{2D}}\right)}{\cosh\left(2\frac{\delta\sqrt{\rho + 1/\tau}}{\sqrt{2D}}\right) - \cos\left(2\frac{\delta\sqrt{\rho - 1/\tau}}{\sqrt{2D}}\right)}. \tag{4.119}$$

Despite the inconvenience of the resulting expression, the Nyquist diagrams for the Gerischer finite-length diffusion element are not very different from the corresponding diagrams for semi-space, which confirms the graph, shown in Fig. 4.70.

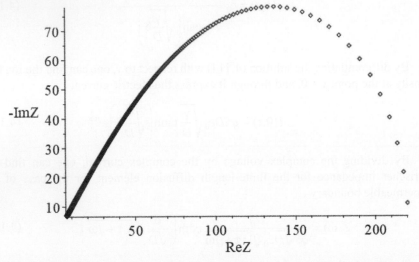

Figure 4.70. Nyquist diagram for the Gerischer impedance of a finite-length diffusion element with impermeable boundary. The calculations carried out for frequency range 1 Hz–5 kHz and parameters $A_W = 1562.5\,\Omega\cdot\mathrm{s}^{-1/2}$, $\tau = 0.01\,\mathrm{s}$, $\delta = 1\,\mu\mathrm{m}$

Gericher impedance for a finite-length diffusion element with transmissive boundary

The boundary-value problem (4.92), (4.93), (4.96), considered here, differs from the previous one only in the last boundary condition, therefore, one could begin its solution right from this stage. Requiring that the solution (4.100) meets the condition (4.96)

$$C_1 e^{\sqrt{\frac{\gamma}{D}}\delta} + C_2 e^{-\sqrt{\frac{\gamma}{D}}\delta} = 0 ,\qquad (4.120)$$

one finds that

$$C_2 = -C_1 e^{2\sqrt{\frac{\gamma}{D}}\delta} .\qquad (4.121)$$

Then the solution of Equation (4.92) takes the form

$$c(x,t) = -2C_1 e^{\sqrt{\frac{\gamma}{D}}\delta} \sinh\left(\sqrt{\frac{\gamma}{D}}(\delta - x)\right) e^{j\omega t} .\qquad (4.122)$$

The unknown constant C_1, expressed from the boundary condition (4.93)

$$2C_1 = -\frac{c_0 e^{-\sqrt{\frac{\gamma}{D}}\delta}}{\sinh\left(\sqrt{\frac{\gamma}{D}}\delta\right)},$$

(4.123)

allows one to obtain the solution of the boundary value problem (4.92), (4.93), (4.95) in the final form

$$c(x,t) = c_0 \frac{\sinh\left(\sqrt{\frac{\gamma}{D}}(\delta - x)\right)}{\sinh\left(\sqrt{\frac{\gamma}{D}}\delta\right)} e^{j\omega t}.$$

(4.124)

This solution allows one to calculate the ion flux density in the plane $x = 0$, and to express the electric current at the same plane

$$I(0,t) = qSDc_0 \sqrt{\frac{\gamma}{D}} \coth\left(\sqrt{\frac{\gamma}{D}}\delta\right) e^{j\omega t}$$

(4.125)

By dividing the complex voltage by the complex current, one can find the Gerischer impedance for the finite-length diffusion element, with an impermeable boundary

$$\hat{Z}(\omega) = \frac{U_0}{qS\sqrt{Dc_0}\sqrt{1/\tau + j\omega}} \tanh\left(\frac{\delta}{\sqrt{D}}\sqrt{1/\tau + j\omega}\right).$$

(4.126)

To separate the real and imaginary parts of Equation (4.126) the above expression for square root of the complex number (4.107) is used. With its substitution into Equation (4.126) it converts to expression

$$\hat{Z}(\omega) = \frac{U_0}{\sqrt{2D}Sqc_0\rho}\left(\sqrt{\rho + \frac{1}{\tau}} - j\sqrt{\rho - \frac{1}{\tau}}\right)\frac{\sinh\left(2\frac{\delta\sqrt{\rho + 1/\tau}}{\sqrt{2D}}\right) + j\sin\left(2\frac{\delta\sqrt{\rho - 1/\tau}}{\sqrt{2D}}\right)}{\cosh\left(2\frac{\delta\sqrt{\rho + 1/\tau}}{\sqrt{2D}}\right) + \cos\left(2\frac{\delta\sqrt{\rho - 1/\tau}}{\sqrt{2D}}\right)},$$

(4.127)

which can be easily divided into real

$$\operatorname{Re}\hat{Z}(\omega) = \frac{A_W}{\rho}\frac{\sqrt{\rho + 1/\tau}\sinh\left(2\frac{\delta\sqrt{\rho + 1/\tau}}{\sqrt{2D}}\right) + \sqrt{\rho - 1/\tau}\sin\left(2\frac{\delta\sqrt{\rho - 1/\tau}}{\sqrt{2D}}\right)}{\cosh\left(2\frac{\delta\sqrt{\rho + 1/\tau}}{\sqrt{2D}}\right) - \cos\left(2\frac{\delta\sqrt{\rho - 1/\tau}}{\sqrt{2D}}\right)}$$

(4.128)

and imaginary

$$\operatorname{Im}\hat{Z}(\omega) = -j\frac{A_W}{\rho}\frac{\sqrt{\rho-1/\tau}\,\sinh\left(2\dfrac{\delta\sqrt{\rho+1/\tau}}{\sqrt{2D}}\right)-\sqrt{\rho+1/\tau}\,\sin\left(2\dfrac{\delta\sqrt{\rho-1/\tau}}{\sqrt{2D}}\right)}{\cosh\left(2\dfrac{\delta\sqrt{\rho+1/\tau}}{\sqrt{2D}}\right)-\cos\left(2\dfrac{\delta\sqrt{\rho-1/\tau}}{\sqrt{2D}}\right)} \qquad (4.129)$$

parts. The system of Equations (4.128), (4.129) is a parametric equation of a Nyquist diagram, shown in (Fig. 4.71).

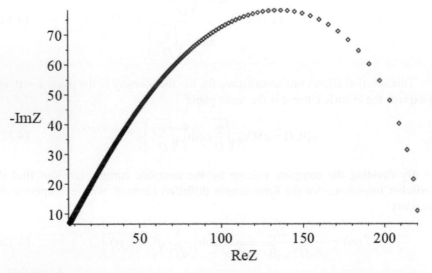

Figure 4.71. Nyquist diagram for the Gerischer impedance of a finite-length diffusion element with transmissive boundary. The calculations carried out for frequency range 1 Hz-5 kHz and parameters $A_W = 1562.5\,\Omega\cdot s^{-1/2}$, $\tau = 0.01$s, $\delta = 1$ μm

Constant phase element

Modeling distributed electrochemical systems within the framework of the structural approach is a rather complicated task. In most cases, the equivalent circuit of an electrochemical system cannot be reduced to a set of ideal elements and their combinations. The matter is that real electrochemical cells are distributed systems in which various processes associated with the transfer, generation and recombination of charge carriers can occur. Above the diffusion mechanism of charge transfer, was considered which in the framework of structural approach is described by distributed elements such as Warburg and Gerischer diffusion elements. But distributed elements, such as solid electrolytes, electrodes etc., can be non-uniform in stoichiometric composition, that can result in a local deviation in the transfer number. Density heterogeneities can cause local deviations of diffusivity. In solid electrolytes, the dielectric permeability of which is significantly lower than that of liquid electrolytes, the charge transfer will be determined both by the concentration gradient and electric field. By virtue of the above circumstances, the experimental value of the impedance will be a value averaged over the volume, and not necessarily coinciding with the impedance of Warburg or Gerischer. Thus some adjustments to these models, considering the non ideality of the diffusion process, should be made.

In a real electrochemical system, in addition to the electrolyte, there are other distributed elements, for example, electrodes and current collectors, in which or at which redox reactions occur, determining the impedance of the system. On closer examination of the electrode surface, it turns out to be as imperfect as the electrolyte itself. The surface contains a large number of defects, such as pits, nicks and protrusions, due to which reactions occur not on the surface, but in a certain near-surface volume. This leads to deviations from solutions of boundary value problems, similar to those discussed above, in which the boundary is an ideal geometric surface. Particles adsorbed by the surface, local inhomogeneities of the charge and also deviations of the stoichiometric composition of the electrolyte result in an unpredictable distribution of the reaction rate over such an interface. Thus the capacity of a double electric layer modeled by an ideal capacity, is not such and when constructing an equivalent circuit, in addition to non ideality of diffusion element, it is necessary to consider the non ideality of the capacity itself. Thus a model is needed that would allow one to consider the deviations of a particular system parameters from the ideal model by appropriate adjustment of its parameters.

The simplest model that formally approximates the impedance of a distributed electrochemical system is the Constant Phase Element (CPE). CPE is an empirical impedance function that, in the simplest case, looks like

$$\hat{Z}_{CPE} = A(j\omega)^{-\alpha}. \tag{4.130}$$

The constant phase element significantly simplifies the process of interpreting experimental data [24]. First of all, this refers to those Nyquist diagrams which have the form of a flattened arc intersecting the real axis at an acute angle.

To model relaxation of polarized dielectrics, the concept of complex dielectric constant $\hat{\varepsilon}(\omega) = \varepsilon''(\omega) + j\varepsilon'(\omega)$ is used. For the convenience of plotting and comparing the Nyquist diagrams, the complex permittivity is usually considered in a normalized and dimensionless form $E(\omega) = (\varepsilon - \varepsilon'_\infty) / (\varepsilon'_0 - \varepsilon'_\infty)$. Within the framework of the Debye model, this function has the form $E(\omega) = 1/(1-j\omega\tau)$. In 1941 Robert Cole and his brother Kenneth Cole, proposed the following of dielectric relaxation model

$$E(\omega) = \frac{1}{1+(-j\omega\tau)^{1-\alpha}}, \tag{4.131}$$

which approximated the flattened arcs in the graphs very well, obtained for a wide range of polar liquids and solids [25]. Equation (4.131) is a modified Debye function for the complex permittivity and reduces to it when $\alpha = 0$. Thus, the Cole-Cole formula was the first model containing the Constant Phase Element (CPE).

For integer values of the parameter $\alpha = -1, 0, 1$, the CPE (4.130) degenerates to ideal elements L, R, C (Fig. 4.72). So, when $\alpha = -1$ one can obtain

$$\hat{Z}_{CPE} = j\omega A, \tag{4.132}$$

i.e., the inductor impedance. Coefficient A in this case has the meaning of inductance $A = L$. When $\alpha = 0$ Equation (4.130) reduces to $\widehat{Z}_{CPE} = A$. Since the impedance of such an element consists only of active resistance, then $A = R$. With $\alpha = 1$ CPE transforms into ideal capacitor with an impedance

$$\widehat{Z}_{CPE} = -j\frac{A}{\omega}, \qquad (4.133)$$

where $A = 1/C$. With $\alpha = 1/2$ CPE is the Warburg impedance

$$\widehat{Z}_{CPE} = \frac{A}{\sqrt{j\omega}} = (1-j)\frac{A}{\sqrt{2\omega}}, \qquad (4.134)$$

and parameter A takes on the meaning of the Warburg coefficient $A = A_W$.

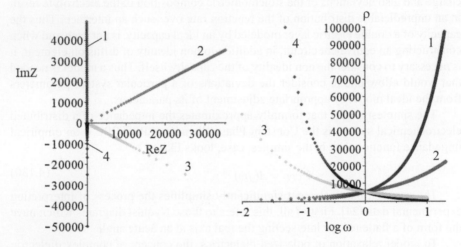

Figure 4.72. Nyquist (left) and Bode (right) diagrams for CPE: 1 for $\alpha = -1$; 2 for $\alpha = -0.5$; 3 for $\alpha = 0.5$; 4 for $\alpha = 1$

Examples of structural models application

Earlier an example of a structural model application to the impedance of a solid electrolyte with platinum electrodes (Fig. 4.61) was considered. It was assumed that in a solid electrolyte, due to its low conductivity, which is three orders of magnitude lower than that of liquid electrolytes, screening of the electric field is not so effective. Therefore, oscillations of lithium ions concentration occur only under the action of an electric field, and their energy is dissipated by the active resistance of electrolyte R_{el}. In fact, the low mobility of lithium ions in a solid electrolyte should result only in the screening field delay with respect to the external potential. Therefore, at low frequencies, charge transfer will be predominantly diffusive. But in LiPON there are two types of charge carriers with significantly different mobilities: mobile lithium ions and immobile charged vacancies. Even if one assumes that vacancies have a mobility similar to the mobility of holes in a semiconductor, then it will take minutes or more to form a vacancy δ-layer at the anode. Therefore, the screening of the field in a solid electrolyte at an alternating voltage with a frequency of more than 1 mHz will always be incomplete. Due to the lack of complete screening, lithium ions will be partially transported by the electric field, and partially by the diffusion flux. This circumstance is considered in the structural model shown in Fig. 4.72, by introducing the Warburg element, connected in parallel with the resistance of the electrolyte.

It was also noted earlier that the small magnitude of leakage current is primarily due to the diffusion mechanism of the current. In the absence of an external source of lithium atoms, a Faraday current is possible only due to the circulation of lithium already contained in the electrolyte. For example in equilibrium, there are two lithium flows, equal in magnitude and opposite in the direction. The first is the ion flux under the action of the electric field in the cathode direction. The second is the diffusion flux of neutral lithium atoms to the anode (Fig. 4.73). To account for the second flux, one more Warburg element, connected in series with the leakage resistance, is added to the leakage current circuit.

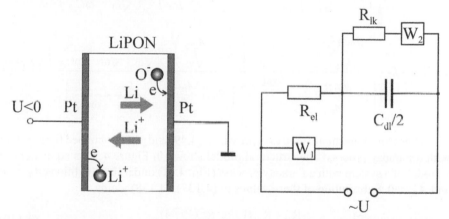

Figure 4.73. Structural model of a solid electrolyte LiPON with platinum electrodes

One feature of the diffusion current that models the Warburg element W_2 should be noted. It comes from the fact that the flow of neutral lithium atoms, in fact, is not a current, because there are no directional movements of electric charges. The charges simply 'disappear' at the cathode and 'appear' at the anode. However, in this structural model, it is not significant, because the system is considered as distributed only at the stage of solving the boundary value problem (4.51)-(4.53). After the impedance expression $\hat{Z}(x)$ is obtained, its value at the point $x = 0$ is assigned to the entire element, and further it is treated as a system with lumped parameters. Obviously, in such an element, there is already no difference between the diffusion and drift processes of charge transfer.

One could calculate the impedance of the structural model shown in Fig. 4.73, using the formulas for the resistance of parallel and serial connection of conductors

$$\hat{Z} = \frac{R_{el}\hat{Z}_{W1}}{R_{el}+\hat{Z}_{W1}} + \frac{\hat{Z}_C(R_{lk}+\hat{Z}_{W2})}{R_{lk}+\hat{Z}_{W2}+\hat{Z}_C}. \qquad (4.135)$$

Substitution in (4.135) the impedances of the Warburg element and the capacitor results in

$$\widehat{Z} = (1-j)\frac{A_{W1}}{\sqrt{\omega}}\frac{R_{el}}{R_{el}+\dfrac{A_{W1}}{\sqrt{\omega}}-j\dfrac{A_{W1}}{\sqrt{\omega}}} - j\frac{2}{\omega C_{dl}}\frac{\widehat{Z}_C\left(R_{lk}+\dfrac{A_{W2}}{\sqrt{\omega}}(1-j)\right)}{R_{lk}+\dfrac{A_{W2}}{\sqrt{\omega}}(1-j)-\dfrac{2j}{\omega C_{dl}}}. \qquad (4.136)$$

The separation of the real and imaginary parts of (4.136) gives the following expressions

$$\operatorname{Re}\widehat{Z} = R_{el}\frac{2+\dfrac{\sqrt{\omega}\,R_{el}}{A_{W1}}}{1+\left(1+\dfrac{\sqrt{\omega}\,R_{el}}{A_{W1}}\right)^2}+\left(\frac{2}{\omega C_{dl}}\right)^2\frac{R_{lk}+\dfrac{A_{W2}}{\sqrt{\omega}}}{\left(R_{lk}+\dfrac{A_{W2}}{\sqrt{\omega}}\right)^2+\left(\dfrac{A_{W2}}{\sqrt{\omega}}+\dfrac{2}{\omega C_{dl}}\right)^2}, \qquad (4.137)$$

$$\operatorname{Im}\widehat{Z} = -j\frac{\dfrac{\sqrt{\omega}\,R_{el}^2}{A_{W1}}}{1+\left(1+\dfrac{\sqrt{\omega}\,R_{el}}{A_{W1}}\right)^2}-j\frac{2}{\omega C_{dl}}\frac{\left(R_{lk}+\dfrac{A_{W2}}{\sqrt{\omega}}\right)^2+\dfrac{A_{W2}}{\sqrt{\omega}}\left(\dfrac{A_{W2}}{\sqrt{\omega}}+\dfrac{2}{\omega C_{dl}}\right)}{\left(R_{lk}+\dfrac{A_{W2}}{\sqrt{\omega}}\right)^2+\left(\dfrac{A_{W2}}{\sqrt{\omega}}+\dfrac{2}{\omega C_{dl}}\right)^2}. \qquad (4.138)$$

These formulas should tend to expressions (4.49) and (4.50) obtained for a system with a Faraday process. The structural model shown in Figure 4.73 is equivalent to a model of a system with a Faraday process (Fig. 4.61) under the conditions $A_{W1} = \infty$ and $A_{W2} = 0$. Substitution of these values in (4.137), (4.138) yields

$$\operatorname{Re}\widehat{Z}\Big|_{\substack{AW1\to\infty\\AW2=0}} = \frac{R_{lk}+R_{el}(1+\omega^2 R_{lk}^2 C_{dl}^2/4)}{1+\omega^2 R_{lk}^2 C_{dl}^2/4}, \qquad (4.139)$$

$$\operatorname{Im}\widehat{Z}\Big|_{\substack{AW1\to\infty\\AW2=0}} = -j\frac{\omega R_{lk}C_{dl}/2}{1+\left(\omega R_{lk}C_{dl}/2\right)^2}R_{lk}, \qquad (4.140)$$

which are identical to (4.49) and (4.50).

Expressions (4.139) and (4.140) allow in plotting the theoretical Nyquist diagrams for the Pt/LiPON/Pt system and compare with the experimental ones. Figure 4.74 shows the experimental (white diamonds) and theoretical (black diamonds) Nyquist diagrams in the frequency range from 1 Hz to 20 kHz. Thus it immediately makes a reservation that the used measurement methods and structural models have frequency limitations. For example, in Fig. 4.74 the experimental curve grows unlimitedly with decreasing frequency, whereas for a real system $\operatorname{Im}\widehat{Z}(\omega)\big|_{\omega=0} = 0$.

It should also be noted that the parameter A_{W1} significantly affects the shape of the curve only in the region $0 < \operatorname{Re}\widehat{Z} \le 50\,\Omega$.

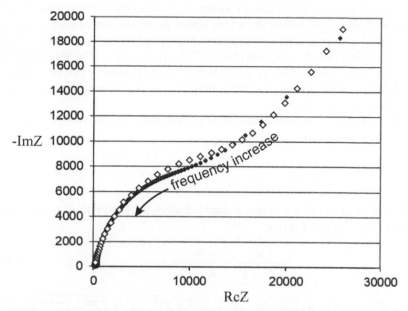

Figure 4.74. Experimental (white diamonds) and calculated (black diamonds) Nyquist diagrams in the frequency range from 1 Hz to 20 kHz with structural model parameters $R_{1k} = 11000\ \Omega$, $C_{dl} = 2 \times 10^{-7}\ F$, $A_{W1} = 10^4\ \Omega \cdot s^{-1/2}$, $A_{W2} = 9 \times 10^4\ \Omega \cdot s^{-1/2}$

According to Fig. 4.74, the structural model shown in Fig. 4.73 satisfactorily describes the variation of a real system impedance in the frequency range from 1 Hz and above and can be considered as an equivalent circuit of the Pt/LiPON/Pt system. The experimental value of the parameter $A_{W2} = 9 \times 10^{-4}\ \Omega \cdot s^{-1/2}$ makes it possible to estimate the concentration of nonequilibrium lithium ions near the electrode. Since the diffusion coefficient of lithium in LiPON is a quantity $\sim 10^{-11}\ cm^2/s$, the area of the electrode is $\sim 1\ cm^2$, and $U_0 = 1$ mV, the amplitude of fluctuations in the concentration of lithium on the surface of the electrode, which is calculated as

$$c_0 \equiv \frac{U_0}{\sqrt{2D}SqA_{W2}}$$, is $\sim 10^{16}\ cm^{-3}$. This is about two orders of magnitude lower than

the equilibrium concentration of lithium.

4.4.3 Electrochemical impedance measurement

Before the advent of modern electronic equipment, the bridge method was the most common method of measuring impedance. Most often, measurements were performed at sound frequencies (20 Hz – 20 kHz) using Wheatstone or Max Wien bridges. The measurement procedure using such a bridge was lengthy and time-consuming, but it allowed in obtaining very accurate results. Despite the fact that these bridges are relatively simple in construction, they allow measuring resistance over a wide range, from several ohms to several megohms, with an error as low as 0.001%. This can explain the fact that at present, with a large arsenal of measurement methods, bridges remain in demand in laboratory practice and in production.

It is customary to divide low-frequency AC bridges into three classes: balanced, unbalanced and quasi-balanced, which differ in adjustment methods and basic metrological properties. By equilibrium of a bridge is meant a state when the potential difference in its diagonal (Fig. 4.75) is equal to zero. The class of balanced bridges includes bridges that, by performing certain operations, can be brought into equilibrium, so, that a change in the voltage of the bridge power supply does not affect the equilibrium state.

The potentials at points A and B, which are calculated as voltage drops at the complex resistances, are

$$U_A(t) = \frac{R_2}{R_1 + R_2} U_0 e^{j\omega t} \text{ and } U_B(t) = \frac{\widehat{Z}_x}{\widehat{Z}_3 + \widehat{Z}_x} U_0 e^{j\omega t}. \tag{4.141}$$

The equilibrium of the bridge is attained when the potentials at points A and B are equal, i.e., $U_A(t) = U_B(t)$ or

$$\frac{\widehat{Z}_x}{\widehat{Z}_3 + \widehat{Z}_x} = \frac{R_2}{R_1 + R_2}, \tag{4.142}$$

where $\widehat{Z}_x = \rho_x e^{i\varphi_x}$, $\widehat{Z}_3 = \rho_3 e^{i\varphi_3}$, $\rho_x = \sqrt{Z_x'^2 + Z_x''^2}$, $\rho_3 = \sqrt{Z_3'^2 + Z_3''^2}$. From this relation, the unknown quantities such as impedance modulus ρ_x of the cell and phase φ_x can be expressed through known parameters

$$\rho_x e^{i\varphi_x} = \frac{R_2}{R_1} \rho_3 e^{i\varphi_3}. \tag{4.143}$$

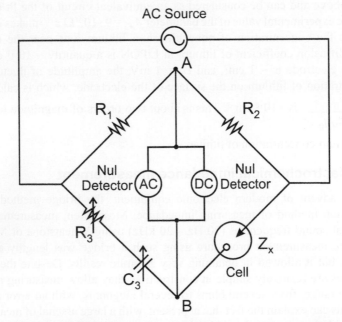

Figure 4.75. Wheatstone balanced bridge for measuring impedance

Equating of the modules and phases in the left and right sides of the expressions (4.143) gives relations

$$\rho_x = \frac{R_2}{R_1}\rho_3,$$ (4.144)

$$\frac{Z_x''}{\rho_x} = \frac{Z_3''}{\rho_3}$$ (4.145)

therefore it is easy to find the imaginary part of the unknown impedance $Z_x'' = (R_2 / R_1)Z_3''$. In view of Equation (4.38), the expression (4.145) reduces to

$$Z_x'' = -\frac{R_2}{R_1 \omega C_3}.$$ (4.146)

From the obtained equations it follows that to balance the bridge, it is necessary to adjust two parameters – the module and the phase. This is achieved by tuning the capacitance C_3 with the help of which the bridge is ac-balanced, and the resistance R_3 by which the bridge is dc-balanced. As an indicator of DC bridge equilibrium an ordinary galvanometer can serve, and as the most sensitive zero indicator for alternating current the Lissajous curves, displayed on the oscilloscope screen, can be used. If the bridge is balanced, then conditions (4.143) and (4.145) are satisfied and with the help of known values R_3 and C_3, the unknowns Z_x', Z_x'', ρ_x, φ_x can be determined for a given frequency.

Carrying out measurements in this way in the required frequency range, one can get an array of data needed to build the Nyquist and Bode diagrams. Further, based on the type of diagrams obtained and the physical model of the studied electrochemical system, it is possible to develop a structural model and calculate the corresponding diagrams for this model. The conformity of the structural model to a real system can be judged by the shape of the diagrams and the values of their parameters. In the case of good agreement between the experimental and theoretical results, it is possible by numerical methods to select model parameters such as, that the deviations of the approximating curve (Nyquist and Bode diagrams) from the experimental one would be minimal. The most common fitting method is the least squares method and its various modifications.

The measurement procedure described above gives an idea of the impedance measurement algorithm, based on the simplest example. Currently, the technique of impedance spectroscopy has gone far ahead and turned into a separate section of circuitry. The frequency range of measurements has also changed significantly, which now extends from tenths of millihertz to tens of megahertz. In modern digital devices, there is no need to balance the bridge with direct and alternating current. The advent of analog-to-digital converters made it possible to process digitized signals and extract the information contained in them using mathematical operations.

Assuming that voltage $U = U_0 \cos(\omega t)$ is applied to the electrochemical cell under the study and the current $I = I_0 \sin(\omega t + \varphi)$ flowing through it is measured. It is required to determine the phase shift between voltage and current. To do this, both signals (current and voltage) are digitized and then multiplied

$$UI = U_0 I_0 \sin(\omega t + \varphi)\cos(\omega t) = \frac{U_0 I_0}{2}\sin(2\omega t + \varphi) + \frac{U_0 I_0}{2}\sin(\varphi) \, . \qquad (4.147)$$

As a result of multiplication, two signals are obtained: a harmonic signal of double frequency and a constant signal equal to the sine of the phase shift.

In analog circuits, low-frequency filters are used to get rid of vibrations at double frequency. In digital signal processing it is much more convenient to use the averaging operation

$$\overline{UI} = \frac{1}{\Delta t}\int_0^{\Delta t}\frac{U_0 I_0}{2}\sin(2\omega t + \varphi)dt + \frac{1}{\Delta t}\int_0^{\Delta t}\frac{U_0 I_0}{2}\sin(\varphi)dt. \qquad (4.148)$$

Under the condition $\Delta t \gg T$ where T is the oscillation period, the first term in the expression (4.148) will be zero, and the more precisely, the more Δt

$$\overline{UI} = \frac{U_0 I_0}{2}\sin(\varphi). \qquad (4.149)$$

Thus by isolating and measuring a constant signal, it is possible to determine the phase of the current as

$$\varphi = \arcsin\left(2\frac{\overline{UI}}{U_0 I_0}\right). \qquad (4.150)$$

To calculate the impedance, it remains only to determine the amplitude of the current, which is obtained by time-averaging of the AC module during $\Delta t \gg T$

$$\overline{I} = \frac{I_0}{\Delta t}\int_0^{\Delta t}|\sin(\omega t + \varphi)|dt \approx \frac{2 I_0}{T}\int_0^{T/2}|\sin(\omega t)|dt = \frac{2 I_0}{\pi}. \qquad (4.151)$$

This relation contains an approximate equality, which holds more precisely, the longer the time for which the averaging takes place.

In some measuring circuits, both signals — voltage and current — are processed in a similar way. They are multiplied by the signal of the clock generator, which has a rectangular shape. Such signal processing makes it possible to increase the accuracy of measuring the phase difference and significantly expand the measurement range. The working range of modern lock-in amplifiers, and, accordingly, impedance spectrometers, lies in the frequency range from 0.5 Hz to 10^5 Hz. Moreover, such spectrometers are equipped with a processor, the measurement procedure is fully automated, and the measurement error does not exceed 0.1–0.2%. There are other options for digital implementation of measurements of impedance spectra, which can be found in [26, 27].

References

1. Goldstein, J. 2003. *Scanning Electron Microscopy and X-Ray Microanalysis*. Kluwer Academic/Plenum Publishers, New York.

2. Bruining, H. 1954. *Physics and Application of the Secondary Electron Process.* Pergamon, London.
3. Mironenko, A.A., Rudy, A.S., Skundin, A.M., Kulova, T.L., Pukhov, D.E. 2017. The method of specific capacity control in negative electrode of lithium-ion battery. RF Patent No. 2621321, 02.06.2017 (in Russian).
4. Barkla, Charles G. 1911. The spectra of the fluorescent röntgen radiations. *The London, Edinburgh, and Dublin Philosophical Magazine and Journal of Science* 396–412.
5. Henry Moseley. 1913. The high-frequency spectra of the elements. *The London, Edinburgh, and Dublin Philosophical Magazine and Journal of Science.* Sixth Series 26(156): 1024–34. London: Taylor and Francis.
6. Manne Siegbahn. 1916. Relations between the K and L series of the high-frequency spectra. *Nature* 676.
7. Emilio Gatti and Pavel Rehak. 1983. Semiconductor drift chamber – An application of a novel charge transport scheme. Presented at 2nd Pisa Meeting on Advanced Detectors, Grosetto, Italy, June 3-7.
8. Lechner, P., Fiorini, C., Hartmann, R., Kemmer, J., Krause, N., Leutenegger, P., Longoni, A., Soltau, H., Stötter, D., Stötter, R., Strüder, L., Weber, U. 1999. Silicon drift detectors for high count rate X-ray spectroscopy at room temperature. 11th International Workshop on Room Temperature Semiconductor X- and Gamma-Ray Detectors and Associated Electronics. Vienna, October 11-15.
9. Bragg W.L. 1914. The diffraction of short electromagnetic waves by a crystal. *Proceedings of the Cambridge Philosophical Society* 17: 43.
10. Wulff G. 1913. Über die Krystallröntgenogramme. *Physikalische Zeitschrift* 14: 217–20.
11. Berdnikov, A.E., Gerashchenko, V.N., Gusev, V.N., Kulova, T.L., Metlitskaya, A.V., Mironenko, A.A., Rudyi, A.S., Skundin, A.M. 2013. A silicon-containing nanocomposite for a thin-film lithium-ion battery. *Technical Physics Letters* 39: 350–2. DOI: 10.1134/S1063785013040032
12. Kulova, T.L., Skundin, A.M., Andreev, V.N., Gryzlov, D.Yu., Mironenko, A.A., Rudyi, A.S., Gusev, V.N., Naumov, V.V. 2015. Cyclic voltammetry studies of silicon–aluminum thin film electrodes synthesized in the presence of oxygen. *Russian Journal of Electrochemistry* 51: 1157–61. DOI: 10.7868/S0424857015120099
13. Kulova, T.L., Mironenko, A.A., Skundin, A.M., Rudy, A.S., Naumov, V.V., Pukhov, D.E. 2016. Study of silicon composite for negative electrode of lithium-ion battery. *Int. J. Electrochem. Sci.* 11: 1370–81.
14. Mironenko, A.A., Fedorov, I.S., Rudy, A.S., Andreev, V.N., Gryzlov, Yu.D., Kulova, T.L., Skundin, A.M. 2019. Charge–discharge performances of the Si–O–Al electrodes. *Monatshefte fur Chemie – Chemical Monthly* 150: 1753–9. DOI 10.1007/s00706-019-02497-1.
15. Airapetov, A.A., Vasiliev, S.V., Kulova, T.L., Lebedev, M.E., Metlitskaya, A.V., Mironenko, A.A., Nikol'skaya, N.F., Odinokov, V.V., Pavlov, G.Ya., Pukhov, D.E., Rudyi, A.S., Skundin, A.M., Sologub, V.A., Fedorov, I.S., Churilov, A.B. 2016. Thin film negative electrode based on silicon composite for lithium-ion batteries. *Russian Microelectronics* 45: 285–91. DOI: 10.1134/S1063739716030021.
16. Wriedt, H.A. 1989. The O-V (Oxygen–Vanadium) system. *Bulletin of Alloy Phase Diagram* 10: 271–7.
17. Stoinov, Z.B., Grafov, B.M., Savova-Stoinova, B.S., Elkin, V.V. 1991. Elektrokhimicheskii impedans (Electrochemical Impedance). Moscow, Nauka Publ. 336 pp. (in Russian).
18. Rudyi, A.S., Vasil'ev, S.V., Lebedev, M.E., Metlitskaya, A.V., Mironenko, A.A., Naumov V.V., Novozhilova A.V., Fedorov I.S., Churilov A.B. 2017. An experimental

examination of thin films of lithium phosphorus oxynitride (a solid electrolyte). *Technical Physics Letters* 43(6): 1–4.

19. Warburg, E. 1899. Uber das Verhalten Sogenannter Unpolarisierbarer Electroden Gegen Wechselstrom. *Ann. Phys. Chem.* 67: 493.

20. Boukamp, B.A., Bouwmeester, H.J.M. 2003. Interpretation of the Gerischer impedance in solid state ionics. *Solid State Ionics* 157: 29–33.

21. Gerischer, H. 1951. Alternating-current polarization of electrodes with a potential determining step for equilibrium potential. *Z. Phys. Chem.* 198: 216.

22. Sluyters-Rehbach, M., Sluyters, J.H. 1970. *In*: A.J. Bard (Ed.), Electroanalytical Chemistry, vol. 4, Marcel Dekker, New York, p. 68.

23. Sluyters-Rehbach, M., Sluyters, J.H. 1984. *In*: E. Yeager, et al. (Eds.), Comprehensive Treatise of Electrochemistry, vol. 9, Plenum, New York, p. 274.

24. Buyanova, E.S., Emel'yanova, Yu.V. 2008. Impedansnaya spektroskopiya electroliticheskikh materialov: uchebnoe posobie (Impedance spectroscopy of electrolytic materials: textbook). Yekaterinburg, UrSU Publ. 70 p. (in Russian).

25. Cole, K.S., Cole. R.H. 1941. Dispersion and absorption in dielectrics – I: Alternating current characteristics. *J. Chem. Phys.* 9(4): 341–52.

26. Andrzej Lasia. 2014. *Electrochemical Impedance Spectroscopy and Its Applications*. Springer New York, Heidelberg, Dordrecht, London, 367 p.

27. Evgenij Barsoukov, Ross, J. Macdonald (eds). 2005. *Impedance Spectroscopy Theory, Experiment, and Applications*. Second Edition. John Wiley & Sons, Inc. Publication. 595 pp.

Conclusion

Summing up the above description of electrochemical systems, materials and technologies used for the development and manufacture of All-Solid-State Thin-Film Lithium-Ion Batteries (SSLIBs), as well as methods for their diagnosis, it is necessary to once again refer to the specifics of this area of research and development. The history of solid-state lithium-ion batteries is comparable in duration to the history of lithium-ion batteries with liquid electrolyte. The starting points could be considered in 1993, when J. Bates and his colleagues synthesized LiPON solid electrolyte, or 1994, when they received a patent for 'Thin Film Battery and Method for Making The Same'. This solid electrolyte is now used in almost all industrial SSLIBs, as it is easy to obtain and reliable in operation. However, the progress that has been made in the field of SSLIBs has been rather slow and the successes here were not as noticeable as in the field of lithium-ion batteries with liquid electrolyte. Although in terms of indicators such as cyclability, rate-capability and the Coulomb efficiency SSLIBs are superior to traditional lithium ion batteries, they are inferior to them in a specific capacity. The main reason is the low conductivity of the solid electrolyte, a significantly smaller effective interface area and the presence of potential barriers at the metal-semiconductor interface. The only way to overcome these problems is the development of 3D SSLIBs, which requires the efforts of research teams of specialists in the fields of electrochemistry, semiconductor physics, electronics and circuitry, mathematical modeling, thin-film and MEMS technologies.

Lithium-ion batteries present the most prospective kind of rechargeable batteries, featuring the highest energy density, reasonable cycle life, quite wide temperature ranges etc. At present such batteries have become the main power sources for portable electronics, including mobile phones and different laptops. In turn, all-solid-state thin-film lithium-ion batteries present some special but rather an important class of the batteries. These batteries provide functioning of microelectronic devices, especially battery-powered Integrated Circuit (IC) cards (smart-cards), radio-frequency identification (RFID) tags, smart watches, implantable medical devices, remote microsensors and transmitters, Internet of Things (IoT) systems and various other wireless devices. Salient feature of all-solid-state thin-film lithium-ion batteries consists in the original technology of their manufacturing. In fact, this technology is a kind of bridge between traditional 'battery' technology and modern technology of semiconductor devices. This feature also affects the design of all-solid-state thin-film lithium-ion batteries. Ideally the battery and the electronic device it powers should be functionally integrated with maximum efficiency and voltage control.

The key issue in the technology of all-solid-state thin-film lithium-ion batteries is the correct choice of functional materials, specifically electrodes active materials and electrolytes. In principle, the electrodes active material for SSLIBs do not differ from materials of traditional lithium-ion batteries, with only the exclusion of, lithium metal that can be used in negative electrode of SSLIBs, unlike ordinary lithium-ion batteries. Indeed, lithium metal provides the maximal specific capacity of a negative electrode, and simultaneously the highest discharge voltage. At the same time, when metallic lithium comes into contact with a solid electrolyte, the problems of dendritic formation and encapsulation (inherent to the ordinary lithium-ion batteries) do not seem to appear. Among other materials of the negative electrodes the elements of the 4th group of the Periodic Table (carbon, silicon, germanium, tin) should be mentioned. All these materials although inferior to lithium metal in terms of specific capacity, are quite correct for SSLIBs. It is worth noting that negative electrodes based on these materials can be manufactured with CVD and PVD techniques, including such popular methods as DC and RF magnetron sputtering. Alternative prospective material for the negative electrode of all-solid-state thin-film lithium-ion batteries is lithium titanate, and especially lithium titanate doped with a variety of foreign cations. This material features by constancy of discharge potential, and provides the highest cycle-life and the excellent rate-capability.

As to functional materials for positive electrodes, the first place is occupied by traditional lithiated cobalt oxide despite the various drawbacks inherent to this material. Along with lithiated cobalt oxide other functional materials must be mentioned, first of all lithium-iron phosphate and the family of vanadium oxides. All these materials possess certain advantages and disadvantages, and their further improvement presents a difficult task for modern investigators.

The main advantage of fully solid-state lithium-ion batteries — their small size — is achieved by reducing the thickness of the functional layers of the battery to 1–5 microns. For the deposition of such layers, Physical Vapor Deposition (PVD) in particular magnetron sputtering methods are used. A detailed description of these methods is given in Chapter 3 of the monograph. Special emphasis is placed on the method of RF magnetron sputtering, which is universal and allows to deposit almost all layers of SSLIB except lithium. A feature of metallic lithium is that in a plasma of a magnetron discharge it is explosively vaporized rather than sputtered. Therefore, for the lithium deposition in systems with a lithium negative electrode, thermal evaporation methods are used. In addition, thermal evaporation can be used for depositing metal current collectors. The information presented in the monograph on the principles of operation of direct and alternating current evaporators and magnetrons will be useful for understanding the specifics of thin-film functional layers deposition, especially nanocomposite electrodes, which are deposited by simultaneously sputtering two or more targets.

As methods for SSLIBs diagnostics, in the monograph are considered by those that are most often used in lithium-ion batteries R&D and manufacture. These are Scanning Electron Microscopy (SEM), X-ray microanalysis, X-ray structural and phase analysis and impedance spectroscopy. Scanning electron microscopy is one of the most informative methods of thin-film analysis, providing a wide range of studies from express analysis to R&D. SEM is indispensable at the stage of SSLIB development for visual control of functional layers, especially of such parameters as:

- layer thickness;
- surface and cleavage morphology;
- adhesion;
- local integrity;
- porosity.

Due to the large depth of the field, the SEM has the ability to two-dimensional visualization of a large area of the sample, which allows the use of SEM as a non-destructive method for detecting defects at the stages of output or intermediate control. The combination of the capabilities of electron microscopy and X-ray microanalysis within one device expands the scope of application of scanning electron microscopes, making them irreplaceable for comprehensive studies of the structure and composition of the functional layers of solid-state batteries.

X-ray spectral microanalysis is required at almost all the stages of thin-film SSLBs development. Such an analysis makes sense if it is carried out simultaneously with the study of the structure, phase composition and morphology of the functional layers. The ability to obtain SEM images of films and data on their elemental composition in one measurement significantly saves time and simplifies the measurement procedure. The role of X-ray spectral microanalysis is especially important when elaborating the technology for the formation of SSLIB functional layers by PVD methods. If, during magnetron sputtering of a target, its material is transferred onto the substrate practically without changing the percentage ratio between the elements, then when depositing films in a gas mixture, the composition of the film can vary greatly depending on the gas flow. For example, for a solid electrolyte LiPON, nitrogen concentration is a critical parameter determining its ionic conductivity. The WDS and EDS methods allow one to determine it with an accuracy sufficient to obtain films with the required characteristics. For the Si-O-Al nanocomposite the ratio of all three components is important, since it determines the conductivity and stability of the electrode. Given the fact that silicon and aluminum are deposited from two different targets when the magnetron power changes according to a certain algorithm, and moreover in an oxygen flow, it is easy to understand how important it is to control the elemental composition of the resulting nanocomposite. The method of X-ray spectral microanalysis allows one to study the elemental composition of the nanocomposite with an accuracy, acceptable for solving technological problems.

Additional opportunities for the technological purposes are provided by the mapping method, with which one can control the distribution of elements. Moreover, electron beam, focused into a fairly small spot, allows one to study not only the lateral distribution of elements, but also the depth distribution. To do this, it is sufficient to prepare an angle-lap or simply cleavage of the SSLIB and probe it by electron beam. Usually, for such measurements the method of secondary-ion mass spectrometry is used, for which the study of the distribution at such depths is a complex and lengthy process. The mapping method can also be applied to study lithium diffusion and migration processes. Although lithium itself cannot be detected by X-ray microanalysis methods, one can see the distribution of the products of its reaction over the surface of the film or its cleavage. Thus, X-ray microanalysis in both EDS and WDS versions is an indispensable research tool in developing SSLIBs.

Characteristics of materials of solid-state lithium-ion batteries, such as capacity, conductivity and stability, are very sensitive to their crystal structure. So silicon, stable in an amorphous state and showing a capacity of up to 3000 mA·h/g, in a crystalline state instantly breaks down at a high degree of lithiation. Higher vanadium oxides V_2O_5 and V_3O_7, which have the largest capacity in a crystalline state, amorphize during charge-discharge cycling and partially loss their capacity. The ionic conductivity of crystalline solid LiPON electrolyte can be an order of magnitude lower than the conductivity of an amorphous electrolyte. From these and other examples it can be seen that the X-ray phase analysis of the functional layers of SSLIB at the stage of their development and production is an urgent task. The described above theoretical foundations and instruments for implementation of X-ray diffraction and phase analysis as well as X-ray microanalysis may be useful when getting acquainted with methods itself and their capabilities as applied to all-solid-state battery materials.

The main difference in SSLIBs and conventional lithium-ion batteries functioning is that lithium transfer processes in a solid electrolyte occur not only through diffusion, but also due to the drift in the electric field. Solid electrolyte is a heterogeneous system in which the mechanisms of energy and momentum dissipation of ions differ from the mechanisms of dissipation in homogeneous liquid electrolytes. The processes of generation and recombination (capture of lithium ions on charged vacancies) create additional resistance to charge transfer. In addition, the effective interface area of the solid electrolyte with electrode is significantly less than in a battery with liquid electrolyte. Lithium ions, accumulating at the interfaces, create a gradient of lithium concentration and, as a result, a back diffusion flow occurs. All these processes are the causes of additional internal resistance and reduction of the Coulomb effectiveness of the battery. The most appropriate method for their investigation is electrochemical impedance spectroscopy.

In contrast to measurements at a direct current or alternating current of the constant frequency, impedance spectroscopy gives a more complete picture of the transport phenomenon, displayed in the form of Nyquist or Bode diagrams. Voltammograms representing the response of the electrochemical system to an external electric field in the range of infra-low frequencies are a significant addition to these diagrams. Interpretation of the Nyquist and Bode diagrams using equivalent circuits makes it possible to determine the magnitude of the active resistance and to establish which particular scattering mechanisms make the main contribution to the active resistance of the battery. Using Nyquist diagrams, one can determine the prevailing charge transfer mechanism, evaluate such parameters of charge carriers as concentration and mobility.

Subject Index

About the Authors

D.Sc. Tatiana Kulova is a prominent scientist in the field of chemical power sources, especially in lithium-ion batteries She is head of Laboratory of processes in batteries in Frumkin Institute of Physical Chemistry and Electrochemistry. She is author of more than 250 papers in scientific journals. Simultaneously she is Professor in National Research University (Moscow Power Engineering Institute). She is a member of the International Society of Electrochemistry.

M.Sc. Alexander Mironenko is the chief technologist of the Yaroslavl branch of the K.A. Valiev Physical-Technological Institute of Russian Academy of Sciences with forty years of experience in the field of micro- and nanoelectronic technologies. Since 2009 he has been developing technologies for the deposition of functional layers of all-solid-state thin-film lithium-ion batteries. For the past 10 years, he has been leading a group of young scientists, working in the field of battery technology and diagnostic methods for materials for lithium-ion batteries. Author of over 40 publications in scientific journals and monographs.

Prof. Alexander Rudy is a specialist in the field of mathematical modelling, dynamics of nonlinear systems and heat and mass transfer in solids. He has been the head of the Yaroslavl branch of the K.A. Valiev Physical-Technological Institute of Russian Academy of Sciences for more than 10 years. Simultaneously he is the head of the Department of Nanotechnology and Electronics in the Centre for Shared Use of Scientific Equipment "Diagnostics of Micro- and Nanostructures" at the P.G. Demidov Yaroslavl State University. Author of 3 monographs and over 150 publications in scientific journals and 14 patents for inventions.

Prof. Alexander Skundin is an acclaimed scientist in the fields of electrochemistry and power sources. For 20 years he headed the Department of Processes in Chemical Power Sources in Frumkin Institute of Physical Chemistry and Electrochemistry. He is author of more than 300 papers in scientific journals and 5 monographs. He is also Professor in National Research University "Moscow Power Engineering Institute". He is a member of the International Society of Electrochemistry and The Electrochemical Society.

Printed and bound by CPI Group (UK) Ltd, Croydon, CR0 4YY

24/10/2024

01778307-0020